EXTINCT

EXTINCT

ANTON GILL & ALEX WEST

First published in October 2001 by Channel 4 Books, an imprint of Pan Macmillan Ltd,
20 New Wharf Road, London N1 9RR, Basingstoke and Oxford.

Associated companies throughout the world.

www.panmacmillan.com

ISBN 0 7522 6162 2

9 8 7 6 5 4 3 2 1

A CIP catalogue record for this book is available from the British Library.

Designed by seagulls
Printed by Bath Press

This book accompanies the television series *Extinct*,
made by Wall to Wall Television Ltd for Channel 4.
Executive Producer: Jonathan Hewes
Series Producer: Alex West
Directors: Jenny Ash; Russell Barnes

For Ted West (1931-2001)
For showing me the beauty of nature and the wonder of science.
Alex West

ACKNOWLEDGEMENTS

Alex West would like to thank:

Ann Hampsey, Artem, Dan Hillman, Gillian Christie, Jonathan Hewes, Jenny Ash, Kate Dart, Helen Britton, Lucy Hall, Parisa Taghizadeh, Richard Attree, Red Vision, Russell Barnes, Skaramoosh and The Hive.

Anton Gill would like to thank:

Marji Campi, Charlie Carman, Gillian Christie, Katharine Dart, Christine King and Alex West.

Thanks also to the following for their assistance:

MAMMOTH:

Joaquin Arroyo-Cabrales, Dean Fisher, Silvia Gonzales, Gary Haynes, Eileen Johnson, Adrian Lister, Ross MacPhee, Paul Martin.

SABRE-TOOTHED TIGER:

Julio Betancourt, Greg MacDonald, Virginia Naples, Chris Shaw, Blaire van Valkenburgh.

IRISH ELK:

Tony Barnosky, Tim Clutton-Brock, Pete Coxon, Silvia Gonzales, Dale Guthrie, Tom Hayden, Andrew Kitchener, Adrian Lister, Tam Ward.

DODO:

Julian Hume, Andrew Kitchener, Beth Shapiro.

GREAT AUK:

Timothy Birkhead, Joanne Cooper, Donald Croll, Jeremy Gaskell, Brad Millen, Willam A. Montevecchi.

TASMANIAN TIGER:

Mick Archer, Jeremy Austin, Col Bailey, Paddy Berry, Judd Case, Eric Guiler, Menna Jones, Nick Mooney, Robert Paddle, Malcolm Smith, Steve Wroe.

PICTURE CREDITS

CONTENTS

INTRODUCTION

When I was an archaeology student at Sheffield University, the dread of end-of-year exam results was magnified by a cruel but funny ritual, the invention of one of my professors. Over the years, he had seen enough mistake-ridden first-year exam answers to paper his office, so he began to save many of his favourite howlers. His file contained several gems: 'The Romans came to Britain because it was too hot on the continent'; 'In the Neolithic period, people made pots and dropped them for archaeologists to find.'

It was his habit to post each new crop on the notice board after the exam and, agonizingly for us, before the actual results were published. We first-year students would read them, hoping that the combination of late-night cramming, three hours of speed-writing and the shamanistic power of our good luck gonks would have saved us from the humiliation of being singled out for display.

Years later, as I began work on the television series *Extinct*, one of these bloopers came to mind. It read something like: 'Evolution may have happened without scientists noticing.' Having been reading up on the subject, it struck me that as well as being funny, a statement like this was actually true in many ways – especially when it comes to extinction. The more I delved into the causes of extinction, and the reasons why some species die out while others survive, the more I thought that maybe science hadn't really noticed much at all. There are still many unanswered questions about how and why the process of extinction works, and the fundamental role it plays in the history of life on Earth.

We're all familiar with Charles Darwin's theory of natural selection (i.e. that living species evolved over millions of years rather than suddenly being created), but when *The Origin of Species* was published in 1859, it turned the scientific world upside-down. Scientists continue to piece together an unending stream of biological research in an effort to understand exactly how evolution works. All this research is aimed at understanding one thing: speciation – the origin of species.

In contrast, relatively little research has focused on the deaths of species. Although precise figures are impossible to verify, palaeontologists reckon that roughly 40 billion species have evolved since multicellular life began almost a billion years ago. On the planet today, every single living entity – plant, animal, coral, insect, bird and all the rest – can be counted up as approximately 40 million different species. Work it out and it's quickly apparent that at least 99.9 per cent of all the species that have ever lived are now extinct. The fact is that in the long term almost nothing survives.

Moreover, on average, species survive for a mere 4 million years. A salutary thought when one considers the fact that the human family has been evolving for around 3 million years since first appearing in Africa's Rift Valley. There's nothing to suggest that we are any more immune to extinction than any other animal. In fact, recent research shows that our closest relatives, the Neanderthals, may have become extinct as recently as 30,000 years ago.

Charles Darwin, *c*1854.

Sobering stuff. Palaeontologist David Raup of the University of Chicago has written about the psychological impact of this perceived fragility, and how our reaction to it depends on which of two contrasting views of extinction we choose to take. The first is positive and based on the Darwinian view, asserting that extinction is always with us as a result of natural selection, and it's those species poorly designed to compete that will die out. The second is more worrisome because it favours altogether more random destruction:

> *Almost all species in the past failed. If they died out gradually and quietly and if they deserved to die because of some inferiority, then our good feelings… can remain intact. But if they died violently and without having done anything wrong, then our planet would not be such a safe place.'*
> David Raup, *Extinction, Genes or Bad Luck?*

From an optimistic or a pessimistic view, it's clear that extinction plays a profound role in evolution. The history of life on Earth is really the history of death, and the question to ask of any living species is not so much how it evolved, but how it has managed to survive at all. Today, the potentially conflicting answers to this question are among the big issues of evolutionary biology and palaeontology. Indeed, as I rapidly discovered, the study of extinction has been dogged by controversy for almost as long as the idea has been around.

The concept of extinction took over 300 years to develop fully, beginning a long time before Darwin published *The Origin of Species*. Like most scientific advances, the process of understanding didn't occur in a vacuum, and a couple of key geological principles had to be established first:

- That the world was very old, and formed of rocks created by the processes of deposition and erosion.

- That rocks contained fossils, which were the remains of creatures that had lived long ago and, crucially, had not survived as a species.

The story begins in Renaissance Italy… In 1482, Lodovico Sforza, Duke of Milan, employed an artisan to prepare plans for road and tunnel building. The man he chose was Leonardo da Vinci. Ceaselessly inquisitive, da Vinci studied the rocks that were uncovered as routes were dug through the mountains. He began to realize something no one else had – that rocks were formed by the deposition of tiny grains of sediment carried in streams, rivers and oceans. He also understood that water could erode the rocks away again, and eventually carry the sediment back to the sea.

Five hundred years later, this basic concept is one of the cornerstones of geology, but to have figured it out in an age that barely accepted the world might be round was a stroke of genius.

Da Vinci noticed that rocks frequently contained the petrified remains of seashells, which posed the question, how did they get there? Today, we know these remains are fossils: creatures that died, and whose organic components over a long period of time mineralized into stone. But in da Vinci's day, no one could conceive of such an idea. It was assumed they were just stones that happened to look like shells, or the result of the biblical flood. As we shall see, it wouldn't be the last time religion blurred the understanding of what fossils are and what they can tell us about the history of life.

Da Vinci looked at the facts and came up with an obvious and sensible explanation: namely, that the sea shell fossils were in some way actually the bodies of creatures that died long ago. The only explanation for their presence in the mountains was that they must have been buried before the mountains were formed. This was a radical concept. Where mountains now stood there must once have been oceans.

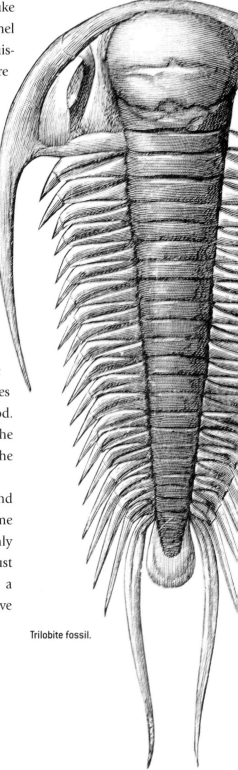

Trilobite fossil.

Since things are much more ancient than letters, it is no marvel if, in our day, no records exist of these seas having covered so many countries… But sufficient for us is the testimony of things created in the salt waters, and found again in high mountains far from the seas.
Leonardo da Vinci

Da Vinci understood that fossils were more than the remains of the biblical flood: they were evidence of life as it was in the distant past. But despite this powerful insight, he never seemed to grasp that fossils often represented extinct species. It was to be another 150 years before the link between fossils and extinction was made and, when it happened, it brought down the wrath of the religious establishment on the scientists.

The vital connection was made by a remarkable scientist called Robert Hooke. Born the son of a country curate in 1635, close to the fossil-rich beds of Freshwater on the Isle of Wight, Hooke can rightly lay claim to the title of England's Leonardo da Vinci. He was a remarkable man – the breadth of his scientific work is astonishing, its importance profound.

Among a catalogue of achievements, Hooke built many flying machines while still a schoolboy; worked on a version of the atomic theory of matter; invented the first truly workable pump to produce a vacuum; used a telescope to make the first accurate observations of lunar craters; and discovered the great spot on the planet Jupiter. He also made 'pendulum watches' by attempting to apply the principle of the pendulum clock to a portable timepiece, hoping this would provide a reliable ship's clock that would enable longitude to be accurately determined.

Hooke's tour de force came in 1665 with the publication of *Micrographica*, his greatest work. He used the microscope to reveal a world never seen before, and *Micrographica* was an instant best-seller, the first popular science book. Samuel Pepys noted in his diary that he stayed up till 2.00am transfixed by its magnified images of a flea, the point of a needle and the edge of a razor blade, to name but a few. Hooke's work also led him to observe that living things were made up of cells, which allowed him to identify the true nature of fossils. Hooke was the first to examine fossils with a microscope and, by comparing them to living organisms, he confirmed that they were also made up of cells – scientific proof that fossils were indeed once living creatures. What's more, he then linked them to the concept of extinction:

> *There have been many other species of creatures in former ages, of which we can find none at present; and that 'tis not unlikely also*

but that there may be divers new kinds now, which have not been
from the beginning.
Robert Hooke, *Discourses on Earthquakes*

Two hundred years before Darwin, Hooke understood that species have appeared and disappeared throughout the history of life on Earth. Extinction was now on the scientific map, but it would be another century and more before this idea was fully accepted.

The majority of people believed in the divine creation of the Earth, and therefore took Hooke's work to be heresy, dismissing extinction altogether. Their reasoning was simple: why would God allow the animals He had created to die out completely? Seeking alternative explanations, some suggested that the 'missing' animals would be found alive when the world had been more fully explored.

It took a third scientist to finally prove extinction beyond question. Georges Cuvier, a statesman as well as a scholar, held high office in revolutionary and Napoleonic France. He was born in 1769, at Montbéliard in the Jura Mountains (the geological formation that would give the Jurassic period its name). As a young naturalist, he came to Paris and worked in the Musée National d'Histoire Naturelle.

Georges Cuvier, engraving by Tardieu.

Cuvier was a truly modern scientist, interested in accurate observation, description and, above all, classification. He divided the animal kingdom into four branches: Vertebrata, Insecta, Vermes (worms) and Radiata (radially symmetrical animals like starfish). His classification system was hierarchical: within each branch, he ranked classes of creature from lowest to highest; the orders in each class could be similarly ranked, and so on down to the species level. Humans sat at the pinnacle of this pyramid of life.

Structure was all-important to Cuvier, and he was a first-class anatomist. He came to believe that all organisms are made up of distinct parts, each designed

Ambroise Tardieu direxit.

to perform its role within the context of the whole creature. Form was related to function. Put simply, all living things are organized to enable them to survive. This may seem obvious to us, but it was a vital advance. Believing that any part of an animal, no matter how fragmentary, showed signs of the form of the whole creature, Cuvier began to reconstruct the physical appearance of animals from their fossilized remains.

While the debate continued to rage around him about whether fossils were really evidence of extinct creatures, Cuvier got on with the detailed observation needed to solve the problem. Using his anatomy skills, he looked at the skeletal remains of animals, including the mammoth. He made detailed observations of modern elephant anatomy and proved not only that the contemporary African and Indian elephants were distinct species, but that the fossil mammoths of Europe and Siberia were different from either living elephant species.

Cuvier's taxonomic classification chart.

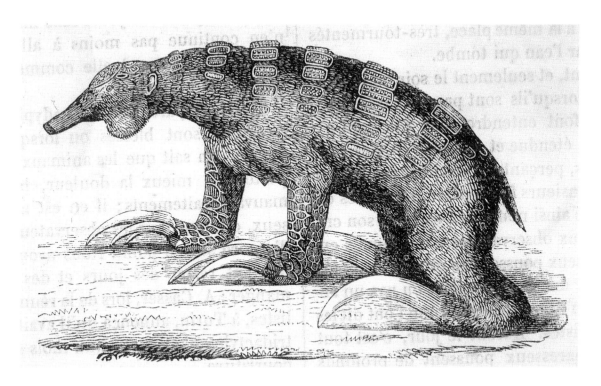

He didn't stop there. In 1812, Cuvier refuted Thomas Molyneaux's 120-year-old idea that the Irish elk was related to the American moose. Cuvier proved it was in fact unlike any other living deer or mammal. He also proved that many other large mammals such as the giant ground sloth and the American mastodon resembled no living species, and his work launched the modern science of vertebrate palaeontology

It was now clear – forty years before the theory of evolution was published – that species had lived and disappeared throughout history. Cuvier didn't actually believe in evolution at all. As unified structures, he felt organisms couldn't change. Since each part was designed to do a specific job within the whole animal, if the parts altered, the creature therefore would not survive.

Cuvier's explanation of why the mammoth and Irish elk had succumbed to extinction held sway for the next fifty years, until Darwin rejected it. By this time it had become known as catastrophism. It held that animal species died out when overcome by some natural disaster. But what kind of disaster?

Many of the fossils Cuvier studied were dug up in alluvial or flood deposits in the valley of the River Seine, close to his home in Paris. So flooding, Cuvier thought, might be one such disaster. This

Megatherium, as restored by Cuvier.

echoed the popular idea of the biblical flood that had wiped out most of life on Earth. But that wasn't the only culprit. After Cuvier died in 1832, his student, Jean Louis Agassiz, observed the glaciers of his native Switzerland. He noticed the marks that ice sheets left on the earth: mounds of debris, U-shaped valleys, and the scratching and smoothing of rocks. Crucially, he also realized that in many places, these signs of glaciation existed where there were no glaciers.

Agassiz used this observation to conclude that a great ice age had once dominated Earth. Giant ice sheets must have covered much of the present-day European and American land masses, destroying all in their wake. This added weight to the view that the planet had been periodically racked by global catastrophes causing the extinction of many creatures. After each catastrophe, new species of animals and plants had then appeared to repopulate the planet.

Catastrophism dominated scientific thinking about extinction in the years before Darwin, conveniently dovetailing science and theology. Animals and birds were created only to be swept away into oblivion, so God's will lay behind the patterns of climate change and natural history observed in the fossil record.

Some followers of Cuvier had suggested that the biblical flood was only the last of many unrecorded catastrophes in Earth's history. Paradoxically, Agassiz's work on glaciers helped to devalue the idea of a biblical deluge. Even so, throughout his years as a scientist, he managed to reconcile his findings with his deep Christian faith. He called ice sheets 'God's great plough'.

Such an accommodation of religious faith and scientific fact was impossible for Darwin. He refuted any religious explanation for the structure and development of life on Earth, including extinction, and was unsatisfied by any idea that luck or fate had anything to do with whether species survived or perished.

Darwin's great insight was a theory explaining the diversity and complexity of nature: natural selection. It was based on, among other things, the concept of gradualism – the idea that changes in nature occur very slowly. The pendulum began to swing away from catastrophism, and Darwin rejected it as a mechanism for extinction. Extinctions of entire species could no longer be blamed on freak occurrences, as Darwin wrote in his seminal work: 'Species and groups of species gradually disappear, one after another, first from one spot, then from another, and finally from the world.'

Darwin believed that extinction was the inevitable outcome for species that failed to win at the battle of life. Survival requires competition and, if adaptation does not equip species to compete, they will be unable to survive. 'Adapt or die' has become the popular phrase to explain this process, but Darwin put it less emotively in *The Origin of Species*:

Satirical cartoon from 1874 – Darwin shows an ape how alike they are.

> *The theory of natural selection is grounded in the belief that each new variety, and ultimately each new species, is produced and maintained by having some advantage over those with which it comes into competition; and the consequent extinction of less favoured forms inevitably follows.*

This idea of the survival of the fittest had such a powerful impact that its influence reached far beyond the boundaries of biology. In the late nineteenth century, so-called 'social Darwinism' gained currency among sociologists and anthropologists. Unfortunately, the idea that the best organisms come to dominate in life became wrapped up with the notion of 'progress', the inexorable development through constant struggle of the most successful forms. Humans were naturally seen as the apogee of this progress, but some humans were seen as less able to progress than others. In time, Darwin's work was misused to develop racist interpretations of human history, with its ultimate expression in the 'science' espoused by the Nazis.

In fact, Darwin steered well clear of the notion of progress or inevitability in evolution – the word 'evolve' is only used once in the entire *Origin of Species*. He preferred the term 'descent with modification'. Darwin was interested in the mechanism of natural selection to explain change and diversity in the natural world.

Yet despite this, he accepted that natural selection does act to improve an organism's chance of survival. By the same token, those species that have become extinct must have had less ability to compete as circumstances changed. Perhaps some environmental shift occurred which they weren't able to cope with, or perhaps another creature adapted to prey on them more efficiently. To paraphrase palaeontologist David Raup, whatever the precise circumstances, extinction is the result of *bad genes*, survival the legacy of *good genes*. Design, not luck, is the key.

When Darwin first published the theory of natural selection, many were violently opposed to his paradigm. Aside from the church and popular opinion, even scientists who had made great strides in understanding biological change and extinction were unconvinced. Jean Louis Agassiz went to his grave still believing that divine creation, not evolution, accounted for life on Earth, and that extinctions were caused by periodic catastrophes.

Some even used Darwin's own idea that creatures had to adapt or die to refute his theory. One animal in particular became a battleground between the anti- and pro-Darwinists: the Irish elk. The elk's magnificent antlers had made it a palaeontological celebrity long before Darwin was published (King Charles II had a set placed in Hampton Court Palace). Why it had such massive antlers (3.5 metres from tip to tip) had long been debated, especially once Cuvier proved it was unlike any other living deer, and was actually extinct. Those opposed to Darwin's theory reasoned that if, according to natural selection, every feature of an animal was designed for improving its chances of survival, then why did the male Irish elk's antlers grow so big, yet still the animal died out? Weren't they just a useless encumbrance that defied Darwinian logic? An example of over-specialization that disproved the theory of natural selection?

It wasn't until the 1970s that the riddle was finally solved, in Darwin's favour, by eminent Harvard palaeontologist Stephen Jay Gould. By measuring antlers and making comparisons, Gould proved that as the beast got bigger, its antler size increased proportionally.

Therefore, the antlers are no bigger than you'd expect in a deer of that size. Their function was to attract a mate and fight off rivals – they were useful. Natural selection does hold true, and they aren't the reason why the Irish elk became extinct.

Gould's study tidied up a biological conundrum hanging over from the nineteenth century, but it did so long after most objections to natural selection had been overcome and Darwin's view of evolution had become scientific orthodoxy. Extinction was now seen as a gradual and inevitable result of evolution at work. A constant level, known as background extinction, could be observed in the fossil record as species slowly lost the fight to survive.

However, new evidence has been uncovered that suggests extinction may not be entirely the result of bad genes, as Darwinist thinking would have it. Ironically, it may well be that luck, not simply efficient design, has a great deal to do with determining which species live and die. Once again the uncertain spectre of the apparently random destruction of species has reared its head and this, above all, makes extinction more than a subject of purely scientific interest. It is relevant for each of us because, quite simply, understanding the nature of extinction challenges our view of ourselves and our place in the world.

The new body of work outlined in detail in this book blows away the cosy feeling (still common despite Darwin's best efforts) that humans are somehow the inevitable inheritors of the mantle of the dominant, best-adapted animal. We have been forced to look again at the history of life, and come face to face with the idea that, in the past, other families of creatures could well have occupied our place at the top of the evolutionary ladder. Could reptiles, birds or any other species theoretically have become masters of the planet while our ancestors remained hiding in the trees?

As the following chapters illustrate, the explanation of the fate of all six animals featured in *Extinct* has undergone major reassessment in recent times. All were perfectly well-designed, successful animals, up to the point of their extinction. The dodo was not the fat, stupid, comical bird we'd like to imagine. The mammoth outlived both the ice age and the efforts of human hunters to kill it. The sabre-toothed tiger successfully survived 100,000 years of glacial conditions, only to perish when the ice age ended. The great auk was just about the best-adapted underwater swimming bird ever to evolve. As we have seen, the Irish elk's antlers can't be implicated in its extinction. And the Tasmanian tiger was doing well enough before the arrival of European settlers on its island home. Finding out what went wrong for these creatures is a compelling mystery which, in each case, suggests the Darwinian picture is an over-simplification.

In recent years, scientists have looked in ever greater detail at extinction. Foremost among them is David Raup. He has examined extinction from a statistical point of view, and his insights are both challenging and entertaining. He is fascinated by the idea that, statistically, most species are abject failures at the game of survival. He set out to shed more light on the question of whether species die out because they have some inherent weakness or are merely victims of chance.

Chance and probability are the domain of the statistician, and in his book *Extinction: Bad Genes or Bad Luck*, Raup neatly uses statistical 'thought experiments' to see if mere chance were in operation in relation to extinction. One such experiment is called 'gambler's ruin'. It draws an analogy between the survival chances of a species and a casino gambler. Starting with a set stake (which can be translated into a number of biological individuals or species), pure chance means that the probability of winning (speciating) or losing (extinction) is exactly 50:50. Of course, this means the gambler may win for a time but, as anyone who has invested in the stock exchange will tell you, past performance is no guarantee of future success.

Sooner or later, the player will lose his original stake and the game will be over. To put it biologically, sooner or later the number of individuals or species will reach zero and extinction will have occurred. This means that if the chance of an organism speciating or dying is initially equal, the depressing conclusion is that extinction is an inevitability.

So the question now becomes: do speciation and extinction actually have a 50:50 probability? Looking at the numbers, this does appear to be the case. The total number of speciations (wins) is almost the same as extinctions (losses). Approximately 40 billion species have lived and died out in the past, and there are 40 million living species today. So the ratio of living to extinct species is 1:1,000. This means that if there have been 40 billion speciations in the past, there have been 39.96 billion extinctions. The figures are almost exactly the same as you'd expect in a game where the chance of winning or losing is 50:50. Luckily for us, and for all the organisms we share the planet with, the slight edge that speciation has over extinction is responsible for the fact there is any life on Earth at all. The chance of living or going extinct isn't exactly 50:50 – it's very, very close, but not absolutely there. The numbers also reveal that extinction is, on the whole, a rare event.

In the casino, the gambler who starts out with a lower stake will reach bankruptcy quicker than one with more cash to bet. Likewise, in biology those species with more individuals, or those groups with more species, will tend to have a better chance of survival. That's why we spend more time trying to get a few golden eagles to breed than a countless multitude of cockroaches. Rarity is a risky business.

There are other factors that give organisms a slight edge. Among these are body size (the smaller the better) and geographical range (the wider the better). Risk of extinction also depends on scale – are we talking about the loss of species, genera, families or entire classes? One example is the dinosaurs. While all species of dinosaur are extinct, the reptiles, the group to which they belong, are still very much alive. But out of all the reptilian species, why did dinosaurs

Fossilized skeleton of a Triceratops dinosaur.

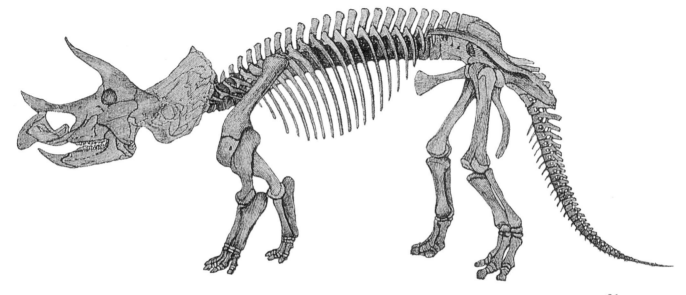

perish and not the others? The results of the second statistical game show that once again chance plays an important part.

It's an altogether more morbid analogy than 'gambler's ruin': the 'field of bullets' scenario. This test measures the likelihood of species being lost by chance in a random killing event – for instance, if a population of creatures were subjected to a hail of machine-gun fire. In fact such an apparently random, yet extremely violent, event now has a name: mass extinction.

Recognizing mass extinctions in the fossil record was one of the greatest scientific discoveries of the twentieth century. This, above all else, has called into question the gradualist, Darwinian view. The fossil record reveals five mass extinctions in the history of the planet, identifiable by the sheer scale of the loss, with many more genera and species succumbing than in other periods, and all in a relatively short space of time, geologically speaking. Mass extinction, in fact, reflects a major collapse of the planetary eco-system, a violent and thankfully rare event. For example, at the time that the dinosaurs disappeared, around 40 per cent of all land-based species also perished. Conversely, the great disappearance of large mammals at the end of the ice age, including the Irish elk, sabre-toothed tiger and mammoth, can't be called a mass extinction, as only a relatively few species of the total were involved.

Subject any group of individuals to the hypothetical field of bullets, and the killing is random. Height, intelligence, weight, eye colour, sex and so on will not help you to survive. Yet there is a pattern. Raup observed that if the population under attack represents an entire biomass, or animal kingdom, as in the case of a mass extinction, the chance of survival decreases towards the level of individual species.

Let us imagine the population under fire are all vertebrates. Raup points out that if, say, 75 per cent of all individuals are wiped out in the hail of bullets, this cannot cause the extinction of all vertebrate animals, since 25 per cent will survive. The chance of vertebrates becoming totally extinct is nil. But if the population of vertebrates is divided into ten species, it's much more likely that one or more of those species will be utterly wiped out.

The field of bullets scenario seems to hold true for mass extinctions. It suggests that in a total eco-system collapse, it's likely many species will be lost purely by chance. It would appear the dinosaurs were one such victim of fate, while mammals were not – but even if the mammals were somehow better adapted to meet the challenges of that particular field of bullets, it wouldn't have mattered to their chances of survival. If mass extinction were to be visited on Earth today, the odds of the mammals as a group surviving would be higher than for any particular species of mammal, including humans. The group would survive, whereas individual species might not. This has profound consequences for the way in which life on Earth has evolved.

Many causes of mass extinctions have been offered, including a drop in sea levels, climate change and volcanic activity. The most radical – and the most famous – implicates a meteor from outer space. This wasn't a serious contender until 1980, when Luis Alvarez and his team of researchers used the discovery of the widespread presence of the element iridium (very rare on Earth) to suggest that impact from an extraterrestrial body had brought it to the planet. Moreover, the iridium layer always sat in geological strata at the boundary between the Cretaceous and Tertiary ages – the exact point where the dinosaurs, and many other species, disappear from the fossil record. The idea that a meteor impact might trigger such a mass extinction suddenly became plausible.

The concept of a huge chunk of ice or rock falling from the sky puts our planet in the centre of a kind of astronomical field of bullets. Raup and his colleague Jack Sepkoski began to compare known impact events with known extinction events. The evidence is inconclusive, but Raup believes that meteor impact is the best mechanism to explain such widespread die-offs. What lives and dies could fundamentally be a matter of chance, an observation that changes our whole view of how life has evolved.

Instead of a steady, inexorable process of evolution, extinction events have created the world and the creatures we see in it today. Every so often, the dominant groups have been wiped out, while others, no better adapted, have been given the opportunity to flourish. Such a flourishing probably does obey Darwinian rules – good design will help a creature survive and compete in normal circumstances. But by definition, adaptations that help survival in normal life can do nothing to help an organism survive events so utterly out of the ordinary.

One might imagine that if rocks fell from the sky regularly, those creatures that could somehow protect themselves might evolve the features that aided their survival to an ever greater degree. But if such an event is so rare that almost none of the species currently in existence has ever had to experience it, there would be no pressure from natural selection to adapt to it.

The stories of the animals explored in *Extinct* seem to reflect this very broad picture in microcosm. While each seemed well enough adapted to normal circumstances, unprecedented events seemed to have overtaken them. The dodo, for instance, having evolved on a single island, had a small population, and was more at risk than if it had been scattered across the globe. Its extinction event, the arrival of a boatload of Dutch sailors, can be seen as akin to a meteor falling from the sky.

As far as scientists can tell, we are now experiencing the sixth major mass extinction to hit our planet. Better understanding of extinction may have a real bearing on whether our own species can adapt and meet the challenges of changing circumstances. Our interest in extinction is timely not merely because it is interesting – it is a matter of survival, of life and death. Our chances of surviving on our own wits may be negligible, but since this mass extinction, unlike any of the others, is of our species' own making, perhaps there's hope that we may act to arrest its more disturbing consequences.

Human civilization began after the last ice age. In a mere 10,000 years, incredible population growth, technological change, and subsequent colonization of every single habitat on the planet have taken place. That makes humans superbly well adapted and, on the face of it, the kind of gambler who might be able to last quite some time in the game, armed as we are with huge numbers and a very widespread geographical distribution.

But as this massive expansion of our species has occurred, it would appear that the rest of the planetary biomass has been enduring a crisis. A mass extinction is simply a question of scale: if a lot of species die out – say, 50 per cent or more – then a mass extinction has

A modern-day endangered species – the giant panda, *Ailuropoda melanoleuca*.

said to have occurred. We are all aware that many plants and animals species are getting rarer, but is this loss enough to be classified as a mass extinction, the point at which survival is decided by chance, not Darwinian adaptation? The evidence looks alarming, and has been summarized by Richard Leakey, famous for uncovering evidence of the earliest human ancestors in his native Kenya.

In his book *The Sixth Extinction*, Leakey gives some alarming figures that suggest that over the next century, some 50 per cent of species may be lost because of the destructive growth in human economic activity. It is impossible to accurately measure exactly what species are becoming extinct; indeed, in the remoter parts of the world, many may be succumbing to oblivion before they are even discovered. For example, in the mountains of Ecuador, the high altitude and rugged terrain create biological islands. In these isolated areas, natural selection has created a range of unique species. In one area, named Centinela, a botanical survey carried out in 1978 discovered ninety new species of plants. Within a few years the area had been turned to farmland and all of these species, which occur nowhere else on the planet, were rendered extinct virtually overnight.

A small loss of habitat can be devastating for a large number of species. It has been estimated that if we end up with just 10 per cent of Earth's tropical rainforest left, at least 50 per cent of the total number of species on the planet will be wiped out. Current estimates reveal that about 2 per cent of tropical rainforest is being destroyed every year, equating to an acre every second.

Even taking the rainforests out of the equation, figures reveal that current extinctions are running at far higher levels than you'd expect under normal background conditions. The figure is easily comparable to any of the big five mass extinctions of the past. It is also happening very, very rapidly, and this is where chance rears its ugly head once more. If 50 per cent of species in the rainforest will be gone in 2050, that gives natural selection no time to allow the

The blue whale,
Balaenoptera musculus,
also on the brink of extinction.

remaining species to adapt and survive. However, even if humankind visits a mass extinction on the planet, it will not mark the end of life on Earth. The planet will recover, albeit slowly, and once again the pattern of life will change. The question is whether we will be there to witness it since, when it comes to mass extinction, luck, not design, is the ultimate arbiter of survival.

And if we did survive, what would we have lost? Our distant ancestors, roaming the steppes and savannahs of the ice age world, saw great herds of mammoth and elk. They heard the roar of the sabre-toothed tiger. In more recent times, explorers stumbled across the dodo and chased the great auk through the surf. Even our grand-parents could have taken a stroll to the zoo to see a Tasmanian tiger. Now, our only hope of experiencing these creatures is through the technology of computer graphics. The television programmes on which this book is based are the closest we can come to a direct phys-ical encounter with animals that have been lost for ever.

It hardly needs pointing out that television is a poor second to the real thing. Will we be artificially re-creating elephants, orchids, penguins, tigers, sparrows, pandas or whales for the television docu-mentaries of the future, or will we be able to see the real things where they belong – in the wild?

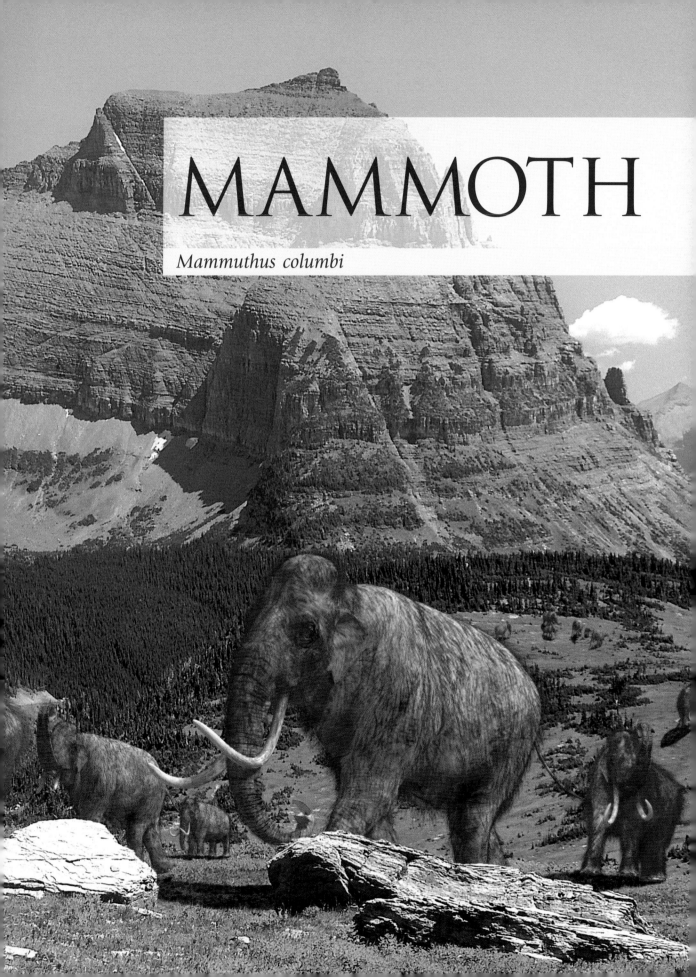

MAMMOTH

Mammuthus columbi

How the mammoth went extinct is a hotly debated subject. One scientist who has made a special study of the problem is Silvia Gonzales, working in the Basin of Mexico. This area is particularly rich in mammoth remains, containing about twenty archaeological sites. In the days of the mammoth the basin was a vast lake, Lake Texcoco, which provided an attractive habitat for the animals. Gonzales has been returning to Mexico, her birthplace, from Liverpool, where she now lives, for many years now. An hour's drive from the centre of Mexico City, in the quiet mountain suburb of Tocuila, a long-term archaeological dig is throwing new light on the mystery of what happened to the mammoth.

In 1996, workmen laying the foundations for a new *cantina* hit something very hard in the ground. Scientists were called in, and in the pit dug by the workmen they discovered the bones of a number of mammoths, including three skulls complete with tusks. It is on these Columbian mammoth remains that Gonzales and her team are now working. Her colleague Joaquin Arroyo-Cabrales has taken detailed measurements of the bones, and through analysing them has been able to determine, by a comparison of tusk size and pelvic bone structure, the sexes of the animals. He calculates that there were seven members in the group – four females and three young males – possibly adolescent calves of some of the females. An infant of perhaps two years of age was also discovered. The matriarch, distinguished by her large skull, is judged by examination of her teeth to have been about fifty years old when she died.

But what brought disaster to this small herd? How did they end up at Tocuila? And how can answers to these questions help us find out how these well-adapted giants went extinct?

On the other side of the world, in what is now Siberia, another mammoth met its fate. The infant had wandered too far away from the herd, perhaps in search of the grass he had begun to eat as he slowly lost his dependence on his mother's milk. Suddenly, he slipped on the edge of the low-lying ground, and slid into a deep pool of soft mud, disguised by the dead grass that had blown across it and stuck on its surface. If he had been a little stronger he might have managed to drag himself to safety, but he was poorly developed

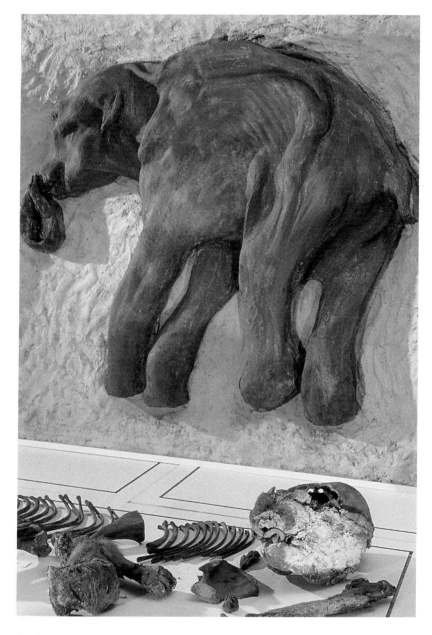

The well-preserved remains of the baby mammoth known as Dima.

for his age – about nine months – and his panic-stricken struggles had dragged him deeper into the mud by the time his mother responded to his squeals and came to the pond's edge. She extended her trunk towards him, but he could not reach it. Other members of the small herd of woolly mammoths hurried up to lend their support, but they could only stand by as helplessly as the mother. Just out of reach, the little mammoth, his struggles already weaker, was sinking deeper.

Soon it was clear to the animals at the edge of the pool that there was nothing to be done, and they drifted off slowly to join the rest of the group. Only the mother lingered a while longer, before she, too, followed the herd as it resumed its endless grass-foraging across the steppe.

This reconstructed scene is based on an informed guess at what must have happened – an event that cannot have been unusual in the lives of mammoths. It took place near what is now the Berelekh River in the western part of the Siberian province of Magadan, around the middle of the last great ice age. His death earned the infant mammoth immortality. The fact that he sank quickly into the mud meant that his body was below the surface as soon as death occurred – which meant that it could not be eaten by scavengers. The fact that he was small and emaciated (he was not a very healthy specimen since his body was thin and his stomach held an abnormally high number of parasites) accelerated the freezing process that soon

followed his demise. It is likely that he died in the early autumn, just before the onset of the extreme winter cold. The ice in the mud drew moisture out of his body, shrivelling and mummifying him, and the following spring new mudflows formed above his grave, burying him deeper and ensuring his preservation, since he would now be out of reach of the changes of the seasons.

And there he lay for about 40,000 years – until June 1977, when a bulldozer driver called Alexei Logachyev, working for a gold-mining concern, uncovered him. He was named Dima after a nearby rivulet. Logachyev found him while looking for gold in the bottoms of ancient rivers that existed in the area before the ice age dried them up. These riverbeds have been covered with waterlogged soil and sediment over the millennia, moving en masse following the thawing of frozen ground. Arctic Siberia is still covered with permafrost – frozen soil to a depth of 500 metres, of which only the first couple of metres thaw during the brief summer – but Dima's remains lay around only 61 degrees north, a relatively low latitude for Siberia. To reach the gold, the miners use high-pressure hoses to disperse the lower permafrost layers, bulldozing the debris away. During this work, Dima was unearthed, still frozen and – uniquely among mammoth finds to date – completely intact, including all his internal organs, though he was damaged by the plough of the bulldozer.

Dima was about 90 centimetres high and 110 centimetres long, and still partially covered with reddish hair. In life, he would have weighed about 110 kilograms. He still had only his milk teeth – the first of six sets mammoths got through during a life of hard chewing – but they were already slightly worn. In his colon remains of plants were found – mammoths were 100 per cent herbivorous and their staple diet was grass. There was earth, and his own hair, in his stomach, indicating panic feeding as he was dying. His near-perfect condition, from the point of view of palaeobiologists, is an indication of the time of year at which he met his end. Most good mammoth remains are of animals that died in the Siberian autumn, to be quick-frozen almost immediately after death. Animals that died and remained on the surface would be subject not only to the elements, but also to the depredations of scavengers – the ancestors of the hyena and the vulture, for example.

Dima's value to the study of mammoths is immense. Soon after his discovery, he was moved to the Academy of Sciences in St Petersburg, preserved by a chemical process. Already a press sensation, he was exhibited around the world. It has been possible to do full analyses of every aspect of his physiology, since even his blood cells remain intact. He has been insured for $12 million.

Dima's discoverer earned himself 1000 roubles and a silver medal from the Academy of Sciences: he was the first and (so far) only recipient of a prize for the discovery of mammoth remains offered since 1860. Originally the prize had been for 100 roubles, if the finder could report a complete skeleton. An additional fifty roubles were to be paid if the Academy was satisfied with the find (that is, it was not a fake). Bowing to the inevitable, the Academy conceded that the finder was at liberty to sell the tusks at his own discretion – 'his' because women would not have been considered eligible in those days. If some of the 'soft' body (the perishable parts) was present, a bonus of 300 roubles would be paid on top. These sums were substantial (but cannot be translated into contemporary money owing to the extreme volatility of the rouble in today's Russian economy) – yet they were still scarcely inducement enough for the peoples of Siberia in the nineteenth and early twentieth centuries.

Under Stalin there was forced labour in the goldfields between the 1930s and the early 1950s, and the practice continued into the 1960s. Other forced-labour camps in Siberia under the Stalinist–Leninist yoke would not have been conducive to the interests of palaeontology. Furthermore, there were strong superstitions among the local Siberian populations, for example the eastern Yakuts and the northern (Arctic) Samoyeds, that it was bad luck to disturb a mammoth's carcass, for reasons that will become clear. Apart from that, one of the early St Petersburg-led expeditions made inordinate and unrewarded demands on local resources for manpower and transport. Intervention from the cities to the west was thereafter resented, and finds were not reported – they would only bring trouble – and the decision not to report them was often taken at the local official level. The problem was exacerbated by a law passed in 1910 forbidding the export of mammoth remains and legally obliging finders to report their discoveries – something that was completely unenforceable.

There are millions of woolly mammoth carcasses in deep-freeze across the millions of square miles that comprise Siberia but, of the thousands that have probably been discovered in the last couple of centuries, barely a dozen have come to the notice of science. Gold-mining companies, if they excavate a mammoth, will in all likelihood bulldoze it over – the arrival of palaeontologists would only provide an irritating delay to the process of making money. The meat of such finds is so fresh (and therefore of inestimable value to science) that local workers as recently as 1983 have boasted of feeding it to their dogs – an enormous irony and a reflection on ignorance, as well as the degradation of the Russian economy, since the recovery of a complete mammoth carcass today would fetch $1 million.

Even so, when the Academy is notified of a find, it is still far too costly to mount an efficient expedition to a site that is usually hundreds of miles from the nearest town. Only international co-operation, organized at lightning speed, could drum up the resources

Mammuthus columbi remains at the Mammoth Site Museum in Hot Springs, South Dakota.

to get to a new carcass in time. Frozen at first, but exposed to the atmosphere about 10–20,000 years or more after its death, the corpse of a woolly mammoth will last only a matter of days. To organize an expedition of all the relevant experts, from archaeologists to palaeontologists to zoologists, is impossible. In such remote areas the dead animal will be torn to pieces by scavengers such as bears or wolverines, or, if discovered in summer as they usually are, decompose fast in the short-lived heat, at the same time as they are devoured by the myriad insects that teem during the season.

The long-dead mammoth is known to us in the present-day elephants of Africa and India – which are themselves fast disappearing thanks to poachers encouraged by the folk-medicine trades of China and Japan, a continuing trade in carved ivory, and the trophy-hunters of the developed world. Elephants share a large number of characteristics with mammoths, and mammoths themselves co-existed in prehistoric times with humankind. Our relationship with the mammoth is both ancient and new, naive and sophisticated. Before considering the origins of the animal, its diversity, and the possibly complementary reasons for its extinction, it will be helpful to look at our own attempts to understand it since it died out.

Mammoths evolved around 1.5 million years ago, developing from the earliest mammoth, *Mammuthus meridionalis*, into three distinct types: the steppe mammoth, the familiar woolly mammoth, and its more southerly ranging cousin, the Columbian mammoth, which lived in the southern and central regions of the present-day USA. The mammoths' range was enormous: dwelling exclusively in the northern hemisphere, it spread throughout Europe and Asia across the land bridge that is now the Bering Strait to Alaska. Eighteen thousand years ago, at the peak of the last ice age, a vast area of land existed which is now under the seas. Not only was the Bering Strait and a great area around it dry, allowing settlement, and migrations across it in both directions, but there was land far to the north of the present north coast of Siberia, and the north coasts of France and Germany. Britain was joined to Europe, and the English Channel and the Southern Bight of the North Sea were dry land.

A huge band of territory that became known as the 'mammoth steppe' ran part of the way round the world between the southerly latitudes of the ice and about 45 degrees north. This was the abode of the mammoth, and, despite the popular image of the snowbound woolly mammoth, the environment was by no means an Arctic waste – rather, it was an enormous grassy prairie, which nevertheless contained a greater botanical variety than the name suggests. Mammoth remains have been found throughout most of Europe and Russia, and the central states of the USA, though the best-known finds have been in Siberia, where the continuing cold climate has preserved carcasses and led to some dramatic discoveries.

Such discoveries worldwide date back many centuries. In medieval Europe any mammoth bones unearthed were assumed to have belonged to an extinct race of titans or dragons. As late as 1600, the Chinese Emperor Ch'ang Shi wrote a bestiary in which he averred that mammoths were giant rodents that lived beneath the earth's surface. Ten years later, when a cache of huge bones was found in England, in Gloucestershire, many believed that they belonged to a race of long-dead giants. The clergy sensed heresy, and one bishop preached a sermon in which he attempted to show that the bones were the remains of an elephant brought over with his army by Julius Caesar.

Even after science had determined that mammoth remains belonged to a vanished race of elephants, those uncovered in southern Europe and especially Italy as late as 1750 were held to have been those of the elephants used by Hannibal. In Russia, similar attributions had also been made: mammoth bones were thought to have belonged to elephants used by Alexander's army during his rapid and dramatic thrust eastwards in the fourth century BC. By the turn of the nineteenth century the great French naturalist, Baron Georges Cuvier, proposed that there had been a race of creatures related to, but distinct from, modern elephants. He published a paper in 1798 demonstrating that mammoths' bones differed from those of elephants. In 1799 the naturalist Johan Blumenbach gave the mammoth a name, *Elephas primigenius* – first-born elephant. He was on the right track, but not quite correct, since elephants are cousins, not descendants, of the mammoth; his term has since been superseded. By that time, remains had been discovered in North America, in the bogland of Big Bone Lick, Kentucky. The bones and tusks unearthed there in fact belonged to a distant cousin of the

mammoth, the mastodon, but in time mammoth remains would also be found in the USA in large quantities.

The word *mammot* first appears in Russian in 1696, probably derived from an Ostiak (Siberian) root, though the Tatar *mamont* may also contribute to the coinage, as well as the Estonian compound of *maa* and *mutt*. The word comes from 'earth' and 'burrower', since the earliest people to see the dead beast lived in the northern reaches of Russia, where it was well preserved by the cold and looked as if it had just died. They believed that the mammoth was some kind of giant mole that lived permanently underground but died if it accidentally breached the earth's surface. Some local cultures attributed malign magical qualities to it (the subterranean world was associated with demons), from which the idea arose that it was bad luck to tamper with the creature's remains.

In those parts of the world that had formerly been part of the mammoths' domain, their tusks were also periodically unearthed. They were subject to a variety of explanations, such as griffins' claws or dragons' horns, but their intrinsic value was also recognized. As early as the fourth century BC the Chinese wrote of the existence of a large animal that they called the *fyn shu*, or 'self-concealing mouse' – an interesting variant on the earth-mole, or burrower, of the Siberians, from whom their information came. They believed it to have a sinuous neck, small eyes, and a stupid nature.

Arab traders of the tenth and eleventh centuries established a brisk trade in mammoth ivory, carrying it from Siberia to Khiva on the shores of the Aral Sea, where carvers worked it into ornaments and combs that were sold at high prices. Mammoth ivory did not find its way to England until 1611, when an adventurer called Jonas Logan bought a tusk from Samoyed tribesmen on the banks of the River Pechora and brought it back to London – no mean feat, as such tusks can weigh around 45 kilograms. Over the centuries a great trade in fossil ivory grew up, and by 1900 the annual market at Yakutsk sold about 20,000 kilograms of it – a testimony to the size of the resource in Siberia, since it represents the combined weight of the tusks of around 220 mammoths.

It has been estimated that during Russia's 300-year occupation of Siberia, the tusks of 45–50,000 mammoths were sold. Mammoth ivory is still used, replacing elephant ivory in an attempt to keep the carving trade alive following the 1979 worldwide ban on the trade in elephant tusks. However, although mammoth ivory is slightly darker than that of the mammoth's living cousins, it is not always possible to differentiate between the two once it is carved into artefacts. The use of mammoth ivory may mask a continuing illicit trade in new objects carved from poached elephant ivory. Today, a mammoth tusk in good condition can fetch the finder around £700.

During the eighteenth century more and more mammoth data, increasingly scientific in focus, came out of Siberia and, in the early years of the nineteenth century, the first major scientific investigation took place.

A few years earlier, in 1799, a Tungus chieftain called Osip Shumakov had noticed on his travels an odd-looking mound of ice near Cape Bykov, at the mouth of the River Lena. Shumakov passed the same way the following year and the year after, by which time thaws had revealed that the mound was in fact a mammoth's carcass – by then one tusk and one side of the body had been revealed. Because of the bad luck associated with anyone who interfered with such things, he hesitated to do anything about his find – a few years earlier another Tungus had taken a tusk from a mammoth and shortly afterwards died with all his family. Shumakov continued to monitor the body over the next couple of years, however, as more and more of it became exposed. By 1803 the mammoth was completely free of its frozen shroud and lay on its side on a sandbank. The profit-motive overcame Shumakov's fear. He cut off the tusks and sold them for fifty roubles.

Three years later, in 1806, an eminent botanist from the Academy of Sciences called Mikhail Adams was travelling in the region and got word of Shumakov's find. He made his way to Cape Bykov but, as no harm had come to Shumakov for his desecration of the carcass, other locals had become emboldened and allowed their dogs to feed on the still (at least to dogs) palatable flesh. Insects and other wild scavengers had also done their work, and, unprotected by a cocoon of ice, the body had in any case started to decompose. Nevertheless, except for one foreleg, the skeleton remained intact, as well as much of the skin and hair. Adams gathered all the fragments together and carefully numbered them before having them transported back to St Petersburg, where the skeleton was reassembled in 1808. He even managed to trace and repurchase the original tusks. The animal had been about 5 metres long and 3 metres high, and had died about 25,000 years previously at an estimated age of forty-five.

Few further discoveries were made during the nineteenth century, though there were many reported sightings, at least one false alarm and one hoax. Local people remained superstitious; and the Yakuts involved in helping Adams had spread the word of the trouble, unexpected labour and haulage work his expedition had put them to. The next expedition did not take place until 1901.

A year earlier, the governor of Yakutsk had sent word to St Petersburg of a mammoth carcass frozen in a cliff by the River Beresovka, well inside the Arctic Circle in the remote north-east of the region. News of the find had reached the governor via a Cossack trader in the settlement of Srednekolymsk, who had bought two tusks from a Lamut tribesman. The Lamut had told the Cossack where the tusks had come from, and that he had not otherwise disturbed the site, for fear of bringing bad luck upon himself.

A major, well-funded expedition was set up under zoologists Otto Herz and Eugen Pfizenmayer. They left St Petersburg by train for Irkutsk on 3 May 1901, and then spent nearly four months in gruelling travelling, during which the team geologist almost died, before finally reaching the mammoth site on 9 September. The scientists were exhausted, but they had to work fast, for winter was already setting in. The mammoth had been partially exposed by a landslide, and most of its internal organs had rotted away, but the bulk of the beast was in good condition – indeed, though most of the exposed head had been eaten to the bone by scavengers, no better sample had ever been recorded.

During the next six weeks, Herz and Pfizenmayer removed about 130 kilograms of flesh from the mammoth's hindquarters, and took even greater samples from the head and abdomen. These were wrapped in hides and refrozen to protect them, and a hut was built over the remains. The dissection also involved taking large samples of skin and hair, and they discovered fresh food in the dead animal's stomach and even in its mouth. Their dogs eagerly ate the flesh not required by the scientists, and they themselves were tempted to try it, though in the end they withstood the pleasure, since the mammoth smelt like 'a badly kept stable heavily blended with [the smell of] offal', as Pfizenmayer remembered.

By 11 October they were ready to leave, facing a freezing journey (temperatures fell to around minus 50 degrees Celsius). They finally got back to St Petersburg in mid-February 1902. The mammoth, later reckoned to have died because it fell into a crevasse from which it could not extricate itself, was judged to have met its end at the age of about thirty-seven, and to have been in the ice for about 30,000 years. Like Dima, the Beresovka mammoth probably died in the autumn, and owed its good state of preservation to the fact that it had been quickly frozen by the rapid onset of winter, with its plummeting temperatures.

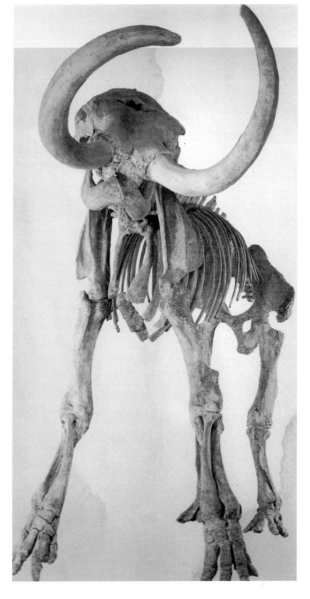

A complete skeleton in the Mammoth Site Museum.

Further mammoth finds were made throughout the twentieth century in Siberia, and they were also made in both western Europe and the New World. The European mammoths were either the woolly species or its antecedent, the steppe mammoth. In North America, woolly mammoths have been found in Alaska, having crossed the land bridge that joined it to Asia when the seas were lower and the relatively shallow Bering Strait was dry. Further south, notably in California, Florida, Kansas and South Dakota, a cousin – a direct descendant of *Mammuthus meridionalis* – has been found: the Columbian mammoth.

Living in a more temperate climate than the woolly mammoth, the Columbian was larger and less hairy. Because of the

climate in their habitat, there have been no frozen remains of perfectly or near-perfectly preserved creatures, but ample fossil evidence has been revealed in the form of bones and tusks. One of the most exciting group of these is in a kind of mammoths' 'graveyard' near the little town of Hot Springs in South Dakota. What probably happened was that, about 26,000 years ago but over the course of several centuries, individuals – most are young males and therefore adventurous loners – fell or slipped into a sinkhole here. A sinkhole is a wide crater fed with water from an underground spring. The water and the vegetation growing on its banks would have made a sinkhole attractive to mammoths, but it has steep, slippery sides and, once in, a mammoth would either drown or, unable get out again, starve. Excavations at Hot Springs over the past quarter century or so have revealed the skeletal remains of about fifty individuals, and estimates suggest that there may be as many as one hundred – not a catastrophic number of deaths considering that they may have occurred over centuries.

It is interesting that at some levels of excavation the remains of woolly mammoths have been found here too, perhaps suggesting a shift in climate at some stage during the ice age when colder climates drove the Columbians further south and allowed the woollies to extend their range further south too, from their normal more northerly latitudes.

Sinkholes were a hazard for mammoths wherever they lived. The myth of elephants' graveyards stems from such places, and from mass deaths through drought or flash floods, as the 'graveyards' are found near dried-out waterholes or river-beds; but for the Columbian mammoths of what is now western California, there was a further danger. They could become trapped in naturally occurring tar pits. A large area of such tar pits, for centuries exploited by man as a source of asphalt, was at Rancho La Brea, off Wilshire Boulevard in downtown Los Angeles. These pits became a rich source of palaeontological material, since not only mammoths but predators became trapped and died there. The predators would be attracted by the helpless mammoths, and in attacking them would themselves become stuck, and die an equally helpless death. (The next chapter presents graphic evidence of such a fate overcoming sabre-toothed tigers, among others.)

None of these natural accidents is enough to account for the extinction of such an adaptable and successful animal as the mammoth. Additionally, a fit animal in the prime of life had no natural predators. But before discussing the reasons for the mammoth's passing, and the possible role man played in it, more should be said about its origins and natural history. It is one of the few ancient creatures we know a good deal about, and we are helped by the living example of its cousin the elephant.

The earliest mammoths, whose habitat stretched from central and southern Europe through into central Russia and the United States, lived on the leaves, bark and fruit of trees, unlike their successors, which were mainly grass-eaters. The climate of the early Pleistocene, about 1.5 million years ago, was mild in the areas these animals occupied, and the landscape more wooded. *Mammuthus meridionalis* looked quite like modern Indian elephants, being relatively hairless, though the single dome of their skull and notably curved tusks marked them out as mammoths. They were also big – about 4 metres tall – with a probable weight of 10 tonnes. But as over the millennia the climate changed, growing ever colder and affecting the landscape, *M. meridionalis* dwindled, and between 750,000 and 500,000 years ago gave way to a form better adapted to the changed conditions: the steppe mammoth, even larger than its ancestor and the biggest of all the mammoths. (Contrary to popular belief, even the largest mammoths were not spectacularly bigger than the modern African elephant, and the woolly mammoth was about the same size.)

The steppe mammoth would have shown the beginnings of a hairy coat, and its diet had started to consist principally of grasses. In turn, as the cold climate of the globe spread further southwards and at a faster rate, the steppe mammoth had given way to the smaller woolly mammoth by about 250,000 years ago. Meanwhile, the Columbian mammoth had developed independently from a branch of the *M. meridionalis* family in the New World, and inhabited a similar prairie-like range, though milder in climate, to its woolly cousin. The later three species had massive, curling tusks, used for fighting, display and foraging. The largest recorded Columbian mammoth tusk is nearly 5 metres long and weighs around 95 kilograms.

At Michigan University, Daniel Fisher has pioneered a method of determining a mammoth's age at death by bisecting its tusks longitudinally. Mammoths continued to grow in size up until the age of forty, and their tusks were formed by the deposition of dentine along the surface of a conical pulp cavity deep within the tusk socket. As successive layers of dentine were added, the whole structure was displaced outwards, until the first layers (of the mammoth's infancy and youth) were pushed outwards and forwards to form the outer

Right and opposite: archaeologists working on a dig at the Mammoth Site Museum.

surface and tip of the mature tusk. The layers of growth are readable in great detail when exposed by bisection, to the extent that periods of greater and lesser growth at various stages in the animal's life can be determined, while thinner and thicker layers, indicative of winters and summers, allow an annual cycle to be established.

Fisher believes that the Columbian mammoth may have been hunted to death, since the evidence provided by his tusk analysis suggests that as the time of their extinction approached, Columbian mammoths reached maturity earlier and grew faster: if food shortage had been the cause of the animals' demise, he argues, one would expect growth rates to decrease and the age of maturity to increase.

Mammoths had prodigious appetites. To keep themselves fuelled, they needed to eat about 225 kilograms of food a day. This meant that, like modern elephants, they slept little, probably spending twenty hours a day foraging and eating. Mammoths are presumed to have had poor eyesight, like modern elephants, and to have relied principally upon their keen senses of smell and touch, transmitted through the trunk. One striking difference is in ear size: a mammoth's ear – particularly that of the woolly mammoth – was far smaller than that of either modern elephant species. This is believed to be because heat would have been lost through a large ear flap. The ear was protected by thick hair. The mammoth's body was covered with two layers of hair: a coarse, long outer guard-hair, acting rather like a weatherproof jacket, and a softer, woollier, insulating fur beneath.

They seem to have been capable of living for up to sixty years, and, like elephants, had a sophisticated, altruistic social structure: mammoths would look after each other and orphaned calves would be fostered, not abandoned. (There is also fossil evidence suggesting that mammoths, like modern elephants, were reluctant to abandon a dead or dying member of their group.) The basic group consisted of perhaps a dozen females, all related, and governed by a mature matriarch. Offspring of both sexes stayed with the group until the age of about ten. Mature males lived solitary lives, coming into contact with females only to mate. Younger males might have formed small corps of their own. Competition for mates was probably intense but, though fights between rival males could be fierce, mortal combat was rare. When it happened – or if called upon to protect young or weak members of the group from predators such as dire-wolves – the tusks became formidable weapons, used for battering and stabbing. The trunk too, with its thousands of muscles, and capable of such delicate actions as plucking sweet herbs or grasses, would have been capable of breaking the back of a wolf. One fossil discovered in Nebraska comprises two skulls locked together by their curling tusks: the combatants had been unable to pull apart, and must have died a slow and terrible death.

It is likely that mating occurred in the summer; with a twenty-two-month gestation period, this would ensure births in the next spring but one. The calves would therefore be able to enjoy relative

MAMMOTH: VITAL STATISTICS

Mammuthus meridionalis (ancestral type)
M. trogontherii (steppe type)
M. primigenius (woolly type)
M. columbi (Columbian type)

Appearance: sloping back, dense fur (woolly mammoths), high, single-domed head, small ears, curved and twisted tusks, short tail, trunk nozzle with one short and one long 'finger' for plucking, holding operations. Short but distinct neck.

Body size: dependent on species; woolly mammoths about 3 metres to the shoulder; Columbian mammoths 3.5–4 metres.

Weight: dependent on species, but around 10 tonnes for a Columbian. Birth weight about 90 kg.

Hair: outer guard hairs long and coarse, six times thicker than human hair, growing to almost a metre long. Underwool thinner, softer and shorter. Colour varies from reddish orange to dark brown. Columbian mammoths less hairy than woollies. Belly hair thick and long, also around ears, under chin and on sides of trunk. Tail had long hair growing from its end. Possibility of a spring moult for a lighter summer coat.

Skin: 2 cm thick, over layer of fat up to 10 cm thick.

Trunk: *c.* 2 metres long.

Penis: *c.* 70 cm long and 9 cm in diameter (Beresovka woolly mammoth).

Note: Mammoths could swim, the trunk curled above the water as a kind of snorkel. Mammoths reached the Californian Channel Islands – the nearest to the mainland, Santa Cruz, is about 40 km away.

warmth for their first few months of life, and their birth at the beginning of the plant-growing season would benefit nursing mothers too. However, this long gestation period resulted in only one calf, which would be dependent on the mother for milk for at least two years (so that a female mammoth would only be able to breed once every four years or thereabouts). Not only was it vital to be able to protect the young, but it was important for only the fittest males to be sires.

Apart from the four main types of mammoth already described, there were dwarf sub-species, which developed on islands off mainland habitats, probably from full-sized ancestors that swam out to them, and then adapted in size to suit the limited food supply. Remains of pygmy Columbian mammoths, a tenth the weight of their cousins and standing only a little more than one metre high, have been found on the Californian Channel Islands, where they lived between 30,000 and 12,000 years ago. On Wrangel Island, off the far north-east coast of Siberia and once joined to the mainland, the remains of diminutive woolly mammoths have been found. Modern science has recently determined that these survived until as recently as 3,700 years ago – far later than had been supposed as the date for the mammoths' extinction.

This means that mammoths were still on earth when the civilization of Ancient Egypt was at its height – the Giza pyramids were built about 4,500 years ago – but of course mammoths in their own habitat had co-existed with humankind far earlier than that. There are many hundreds of cave paintings of mammoths, notably in France, executed between 30,000 and 10,000 years ago, and drawn with an immediacy and fluidity that means that they can only have been taken from life. In various parts of eastern Europe and western Russia, too, a 15,000-year-old culture has been identified which built its huts of animal hides spread over a framework of mammoth bones and tusks.

Neither the paintings nor the huts should imply that mammoth-hunting was extensive, if it existed at all. Given the enormous amount of mammoth remains that still exists in Russia today, the hut-dwellers would have had a plentiful supply of bones as convenient, if heavy, building materials. The bones, incidentally, also provided a source of fuel – there were no great forests to supply wood in the ice age climate at the time in those parts of the world. As for food, for both the cave-dwellers and the hut-dwellers easier meat was available in the form of animals such as reindeer. Interestingly, though mammoths and horses are frequently found in cave art, reindeer are not. Mammoth ivory was worked and carved by ancient peoples, while usefully shaped bones, such as shoulder blades, were fashioned into axes and even, conceivably, percussion instruments.

But by 15,000 years ago, time was running out for the vast majority of mammoths. They were not alone, as across the world most of the very large mammals that had existed in the late Pleistocene epoch died out between 40,000 and 10,000 years ago. In North America, the Columbian mammoth shared its fate with such animals as the giant sloth and the sabre-toothed tiger (as we shall see in the next chapter);

in Eurasia, the woolly rhinoceros and species of giant deer shared extinction with the woolly mammoth.

Where the two species of mammoth are concerned, the woolly mammoth seems to have been extinct in Europe by about 12,000 years ago. It hung on longer in Siberia – much longer in the case of the dwarf group on Wrangel Island. In North America, the Columbian mammoth seems to have hung on until about 10,600 years ago, to judge from data available at sites in Colorado and Kentucky, though new discoveries, described below, suggest a later date. The reasons for the extinction of such an adaptable mammal are complex.

By about the time of the mammoths' extinction, the last ice age was coming to an end and the climate throughout mammoth territories was rapidly becoming warmer and/or wetter. The special, varied grasslands on which the mammoth depended throughout its southern range were in the main rapidly encroached upon by trees, which grew in response to climatic changes favourable to them. Between approximately 13,000 and 9,000 years ago the prairies of western Europe were swallowed up by forest – too short a time for the mammoth to effect any evolutionary adaptation. Pockets of grassland remained, but mammoth groups became isolated from each other. Being slow breeders, they were especially vulnerable to any threat to their population size, and the diminished grassland could not maintain viable populations. Meanwhile, the northerly ranges of the mammoth steppe were giving way to boggy tundra, an equally unviable habitat from a mammoth's perspective.

At the same time, the change of climate in northern hemisphere America was especially beneficial to a very competitive species – man. Man was also developing more sophisticated hunting skills, and this was particularly bad news for the Columbian mammoths. A group of adept hunters, who probably crossed into America by the same Bering Strait land bridge that had allowed the mammoth in, had established itself in what is now the southern part of the United States. This was about 12,000–11,500 years ago, though the most recent research indicates that they may have arrived as early as 13,500 years ago, or even earlier. Called Clovis Man after the site at Clovis, New Mexico, where their culture was identified, they developed a particularly effective flint spearhead – the Clovis point. Spearheads have been found in a small number of fossil mammoth bones. Clovis Man may also have been able to construct pitfall traps deep enough to stop a mammoth in its tracks so that it could be speared to death at leisure. Given their technology, the fact that the arrival of these hunters predates the demise of the mammoth seems to present an irresistible argument.

It has also been suggested that human hunters may have been able to stampede mammoths in whole groups towards their death – over a cliff, for example. This hypothesis does not have many takers, though a possible site for such an occurrence exists on Jersey in the Channel Islands, where a number of mammoths do indeed appear to have tumbled into a ravine. It is just as likely, however, that the animals fell to their deaths by accident, and the local Stone Age human population took what advantage they could of this unexpected windfall of meat.

There does not seem to be quite enough evidence for the theory of 'overkill' – aggressive hunting to extinction – to hold water. There are still, however, some adherents to the idea developed several years ago by Paul Martin of a kind of 'blitzkrieg' to account for the sudden and virtually contemporaneous disappearance of so many large mammals in North America. The blitzkrieg theory argues that the first humans to arrive from Siberia found a kind of Garden of Eden, teeming with four-legged food that had no fear of man, never having met him. Although the animals were large, there was no need to be economical, since there were so many of them: you could just let what you could not eat go to waste and make another killing as soon as you needed to. In such fecund circumstances man could sweep across a continent, moving on as soon as he had depopulated one area – a kind of scorched earth policy, and well in keeping with human behaviour patterns. Martin said that 'the animals totally new to this kind of encounter will vanish'. The killing front would advance perhaps 100 miles in ten years, and soon victim species would be no more. Fossil evidence for such a theory is hard to find, but Martin contends that the action was so fast that any such evidence could only lie in a very thin geological stratum, and in any case carcasses abandoned by hunters and left on the surface would be scavenged very quickly, even the bones disappearing before they could be buried by time.

Many species have been hunted to the verge of extinction, but only in very recent times could highly mechanized and widespread forms of predation by man (coupled with incursions into or pollution of habitats) have led to the destruction of a species: the blue whale and the Bengal tiger would almost certainly be totally extinct by now, for example, were it not for the intervention of powerful conservationist lobbies, and even so their days are probably numbered.

Primitive man had plenty of other easier quarries than the mammoth to hunt, such as deer, and although it is possible that in colder climes he could have butchered a mammoth and then stored its meat in ponds that would freeze in winter, preserving the meat until the following spring, this cannot have been a very widespread activity. This is a theory, however, which has been the subject of a successful practical experiment (using horse meat) by Daniel Fisher of the University of Michigan.

There are few indications that the mammoth was hunted sufficiently actively in Eurasia for 'overkill' to be a possibility there, but in North America Clovis Man's impact may have been more tangible. Even so, human groups were still relatively small and isolated, and not reproducing as fast as they would as humankind became increasingly successful and dominant. Killing a mammoth might have meant getting a lot of food from just one kill, but would it have been worth the effort, when so much of the meat would rot before it could be eaten, or have to be left inaccessible under the ice of a frozen pond throughout the winter? It may be that hunting contributed to the demise of the mammoth, which would have occurred anyway without it, but over a longer period, as a result of the gradual disappearance of the habitat and food sources that suited it. As for a changing climate alone, mammoths could arguably have adapted to that.

In the case of the Tocuila animals investigated by Silvia Gonzales and colleague Joaquin Arroyo-Cabrales, one possible solution immediately presents itself. The Basin of Mexico lies within an actively volcanic area. Analysis of soil deposits reveals that Mount Popocatepetl, the famous volcano only 20 kilometres from Tocuila, was active at the time these mammoths were alive. Is it possible that they suffered a similar fate to the citizens of Pompeii? At first sight, the likelihood of this seems strong. The bones have been well preserved because for thousands of years they have been hermetically sealed in pumice – liquidized volcanic ash that hardened round them. Once the mammoths had been engulfed in the maelstrom they would have been swept along in a tide of volcanic mud until they finally came to rest at Tocuila.

But was it a mudslide that actually killed them? When the bones were studied in their original positions, both from above and from the side, it appeared that although the skulls with their tusks remained intact and one pelvis and leg were still connected, most of the one thousand bones here were scattered and jumbled up. This would not have been the case if the mammoths had been alive when struck by the volcanic storm. Gonzales takes the scattered nature of the bones as clear proof that the animals were in fact dead when their remains were borne away in the eruption. The mudslide, she argues, broke up their decaying carcasses, scattering their bones. Furthermore, from evidence of slight weathering on the bones themselves it is apparent that the animals had probably been dead for two or three

A partial skeleton unearthed at the Mammoth Site Museum.

months before the eruption. So we are back to the question: what did they die of – or what killed them?

As we have seen, climate change played a role in the process. Tests on pollen samples between 13,000 and 11,000 years old show that on the northern American plains, grassland, birch and willow were giving way to oak, maple and elm forest, while in the southwest, rising temperatures were already creating a desert scrub in which little other than cacti grew. Gonzales, however, is not convinced that the climate effect applies to Mexico. Examination of the earth's stratification here reveals periods of drought, for example, which the mammoths of the Basin of Mexico withstood for thousands of years. Lake Texcoco provides a kind of massive oasis in the encroaching desert – an oasis into which the mammoths withdrew, and where, according to Gonzales, they could have survived for even longer than they did.

Enter Clovis Man. As we have seen, he came across the Bering Strait land bridge some 12,000 years ago and rapidly spread south: archaeological sites across North and South America show that he had reached southern Patagonia within a thousand years of his arrival on what is now the Seward Peninsula. Not far from the Tocuila mammoth site is another containing the remains of two more mammoths, and these were found with Clovis-point spears in them, clearly aimed at the animals' vital organs. But the anthropologist Eileen Johnson, who is an expert on the Clovis culture, does not believe in the overkill theory. She believes that while a few mammoths found with Clovis points in their skin were clearly hunted down, there may be other explanations for butchery marks on the bones of other fossil remains.

Johnson has found that while marks on many bones do appear to have been made by some kind of stone tool, as is the case with some of the material at Tocuila, no weapons were found near the site, nor were any Clovis points in the animals. The conclusion Johnson draws is that the animals were cut up by man, but probably only after they were killed. Just as in the case of the Jersey mammoths, primitive man may have simply taken advantage of a supply of meat provided by chance. And though the Clovis culture also made artefacts from mammoth material, there is little evidence that much mammoth was eaten. In Clovis communities, such as that in Meadowcroft, Pennsylvania, food remains discovered include fish and small game such as rabbits. Johnson does not believe that Clovis man hunted the mammoth to extinction, as large-scale hunting simply wouldn't have been worth it. A few mammoths, maybe older, weaker individuals, might have been brought down, perhaps as some kind of macho coming-of-age feat; but it would have taken a lot of concerted effort to hunt mammoths and there is no evidence that Clovis people were at more than a rudimentary level of social organization.

Even the dating of the end of the Columbian mammoth has now been thrown open to question. During her work on the Tocuila site in the year 2000, Silvia Gonzales found an unusually well-preserved rib towards the bottom of the pile of skeletal remains. When she had it radiocarbon-dated she found the result astonishing. It was found to be only 7,000 years old – which means that mammoths were around easily at the time of the first sophisticated civilizations of the Near East. Ancient Jericho, for example, dates to between 8,000 and 6,000 BC. And the Tocuila site has revealed yet another secret. In one jawbone, in which the teeth are still set, there is evidence of distortion in the molars on one side. Gonzales hazards a guess that it is due to some kind of long-term illness, perhaps caused by toxicity, which itself might have come from volcanic ash in the atmosphere, but might also be attributed to some kind of wasting disease.

The disease idea has given rise to a brand-new theory of the mammoths' extinction, at least in North America, which is being championed by Ross MacPhee of the Natural History Museum of New York and his colleague Peter Daszhak. MacPhee has studied extinctions of mammals in various parts of the world, and it is his belief that major plagues could account for sweeping extinctions from about 40,000 years ago in Australia through to those of 10,000 years ago or so in North America.

MacPhee had never been entirely convinced that the mammoths' demise could have been caused either by climate change (and attendant changes in vegetation in North America) or overkill. His reading on diseases such as AIDS and ebola led him to a brand-new possibility: could a disease carried by man have jumped species and become a death sentence for the mammoth? AIDS, which originated in monkeys, was a very striking reminder that some diseases could cross species boundaries with fatal effect. Ebola, with its 90 per cent mortality rate, suggested a disease capable of wiping out an entire species. With these examples firmly fixed in his mind, MacPhee is now testing for the presence of disease in mammoth DNA. Although he is only now in the foothills of his research and has already met with scepticism, many fellow scientists are watching his progress with interest. In Mexico his theory carries a kind of historical poignancy, for it was the diseases that the conquistadors carried with them – measles and influenza – which proved fatal to the Aztecs in the sixteenth century. (The Aztecs had shown the Spaniards a mammoth's femur, as Bernal Diaz del Castillo, the chronicler of Cortés' expedition, reports in *The Conquest of New Spain*.)

However, for a disease to jump from group to group of the same type of animal, let alone across species, there has to be physical proximity. The question is, how close did Clovis Man ever get to the mammoth? Perhaps he did not need to. If there were a disease that would kill mammoths (but not its carrier), there were plenty of other creatures besides man that followed his progress from Siberia southwards into the Americas – rodents, birds and insects, for example. And by the time Clovis Man appeared on the scene it is possible that he had already begun to domesticate the dog. Two species of dog at least had been domesticated in Europe by the late Stone Age, as sites in Denmark and Switzerland attest. Might a disease akin to bubonic plague – something transmitted by epizootic parasites like fleas – have been carried from canine to mammoth? MacPhee postulates a disease with a swift effect, and that makes palaeopathology harder, because long-term illnesses, like arthritis, leave an impression on bone, whereas in a fast-acting disease, lethality comes so rapidly that any pathological effect you might expect to find on bone just does not occur. Which means that Gonzales' deformed mammoth tooth may represent a false trail, at least for MacPhee.

It is not yet clear exactly how diseases were transmitted, but the warmer climate at the end of the ice age would allow new diseases to emerge and new carriers to thrive. A disease would have been passed on more quickly once inside the mammoth herds because population density was high – the seas had risen and covered vast areas of land, while old habitats were unsuitable, and the last mammoths were tightly packed into refuges like the Basin of Mexico. Given that the mammoth hadn't encountered this hypothetical new disease before, it wouldn't have been able to develop antibodies quickly enough to fight it off. It will be hard to find fossil evidence for a fast-moving, lethal disease, for the same reasons as for the blitzkrieg theory, though the taking of DNA samples from mummified remains (as MacPhee proposes) is one way forward.

But there were great losses of species on the planet long before man even arrived – the disappearance of the great dinosaurs is the most famous. In the end, it seems likely that a combination of effects, which may have been in differing proportions to one another at different points around the globe, brought the mammoth to extinction. Some romantics have even suggested that it is not extinct at all, but may still exist somewhere in small remote enclaves among the millions of square miles of the Siberian tundra. If it does, it will have had to adapt yet again to a new habitat, though palaeobotanical work on Wrangel Island, where the last mammoths lived, indicates that the vegetation there hasn't changed much in 8,000 years. Attractive as the possibility is, it seems unlikely, however, that the mammoth will follow the example of the coelacanth, and unexpectedly turn up alive and well, having been believed dead for so long.

THE CONCEPT

No one knows exactly how the current mass extinction may impact on our own chances of survival. But as global biodiversity is lost, there is one price we are already paying. As each species disappears, we lose the chance to directly experience that animal ever again – not something we generally think about before an animal goes extinct.

Today, we are fascinated by dinosaurs and other prehistoric beasts. We watch films, read books and visit museums in an effort to get as close as possible to something we know we can never experience for real. We wanted *Extinct* to bring the viewer as close as possible to six long-lost creatures.

By telling the story of how each one met its fate, we also wanted to investigate the science behind the mysteries and mechanisms of extinction. Essentially, *Extinct* was conceived as two programmes in one: a science documentary and a wildlife film. The idea was to interweave the two forms. What we learned in the documentary would be applied to our wildlife re-creations. Our hope was that the audience would not only get to observe accurate animal behaviour, they would be shown exactly how the experts came to their conclusions. As we

The dodo skeleton
and clay model.

increased our scientific understanding of the demise of each animal, the wildlife film would illustrate the events unfolding before our eyes.

In order for this ambitious project to work, the appearance and behaviour of these six extinct creatures had to be reconstructed as accurately as possible, using the latest scientific research. Choosing which animals to feature from the millions that have become extinct was initially very diffi-cult. The aims were to cover a large time span, from prehistoric to modern day, and to use a mix of birds and animals. The creatures had to have an inherent interest and visual power, and a compelling story, illuminating the most representative processes of extinction. This excluded creatures such as the dinosaurs, which many scientists now believe were wiped out by the impact of a meteorite, a freak event

Model of 'dead dodo' carcass.

Filming the Tasmanian tiger animatronic in Truwunna Wildlife Park.

outside the normal range of experiences that animals can adapt to cope with.

After painstaking research, the species were chosen: three animals that had lived during the last ice age – the Columbian mammoth, sabre-toothed tiger and Irish elk; and three that became extinct much more recently – the dodo, great auk and Tasmanian tiger. While obviously very different from each other, these creatures had one key thing in common: they had rarely, if ever, been re-created for television before, and never to this level of detail.

However familiar they are from exhibits and drawings, no one alive today has ever experienced these creatures in the flesh, with the exception of the Tasmanian tiger, or thylacine. So while fossilized bones, hair and even skin from the other creatures have survived to give us crucial clues, it is still extraordinarily difficult to capture the realism, drama and wonder of a herd of mammoths travelling across grassy steppes; sabre-toothed tigers stalking through ice age forests; Irish elk stags fighting on the windswept tundra.

How do you make a natural history documentary about animals that have been dead for hundreds or thousands of years? Until recently, it was virtually impossible, from a technical point of view. Then Steven Spielberg made the feature film *Jurassic Park* and the world saw how computer animation could re-create lost worlds with a realism never seen before, albeit at enormous cost. Since then, advances in computer graphics technology have meant that television is now invading this new technical frontier, and one result is the *Extinct* series.

The most significant technical advance was the recent development of software that can accurately, and cost-effectively, create animal fur. When *Jurassic Park* was made in the early 1990s, these software advances were still in development, and it was far easier to re-create the leathery scales of reptiles such as dinosaurs. When the final building block was in place, the process of making *Extinct* a reality could begin. It would be another year and many technical and logistical challenges later before work was complete.

Sabre-toothed tiger animatronic head: interior (left) and finished model (right).

SABRE-TOOTHED TIGER

Smilodon fatalis

Ten thousand years ago, the fertile plains of coastal California were home to a great variety of animals. They lived among the grassland that stretched from horizon to horizon, punctuated by woodland and dotted with shrubs, the yellow-ochre of the grasses and the green leaves of the trees dominating the landscape.

Nearly all of those animals were to die out. Among them were Columbian mammoths – just one of a number of mammals that were far larger than any existing today. Some of these mammals belonged to unfamiliar species – the giant ground sloth, for example. Others belonged to species that are still recognizable: there was a giant camel, and also bison – *Bison antiquus*, the most commonly found herbivore in ancient Californian deposits, which was not only larger than the present-day animal but lived in smaller herds. Because of their size, these animals were slower moving than their modern counterparts.

The presence of so many herbivores – including some non-giant species such as the wild horse that roamed North America before the advent of man – inevitably attracted a range of predators. Among these were the dire wolf, *Canis dirus*, the short-faced bear, *Arctodus* – and several cat-like mammals, including the cheetah-like *Miracionyx* and the lion, *Panthera atrox*. But, as the last great ice age drew to a close in what is now the south-western part of the USA, there was no 'big cat' quite so striking, nor so dominant, as the animal now known as the sabre-toothed tiger, *Smilodon fatalis*.

Its popular name is misleading: *Smilodon* was not closely related to the creatures now called tigers. It belonged to a sub-family of the Felidae, called the machairodonts. The earliest groups contained the genera *Proailurus* and *Pseudaelurus*. *Proailurus* is an ancient type, a smallish animal with a skull no greater than 15 centimetres in length. It is thought that the machairodonts split from the group that went on to evolve into the modern family of cats, but continued to evolve along parallel lines. The machairodonts successfully existed over great parts of the world for several millions of years. They appeared on the scene about 12 million years ago, but their previous origins are still unclear. The sabre-tooths' ancestor was an animal known as Megantereon, which was the size of a mountain lion, and lived right across the northern hemisphere, establishing itself in the New World about 5 million years ago.

The New World branch evolved into the cave-dwelling sabre-toothed cats; other branches became extinct. There were several different species of sabre-tooth – the earliest and smallest was *Smilodon gracilis*, which appeared 2.5 million years ago, and is now known mainly from the eastern USA. *Smilodon fatalis* evolved around one million years ago, and ranged as far south as the Pacific coastal area of South America. Two sub-species were separated by the Andes. *Smilodon* differed from its ancestor Megantereon in that it was bigger and had longer canines, among other details.

The 'tiger' is particularly well documented as so many of its remains have been preserved – in an unusually pristine state – in the tar pits of La Brea in Los Angeles, California. More of them have been found there than in the rest of the world put together: about 162,000 bones from at least 2,100 individuals.

Tar pits form when crude oil seeps to the surface through fissures in the Earth's crust; over the millennia, especially in times of drought,

A complete *Smilodon fatalis* skeleton.

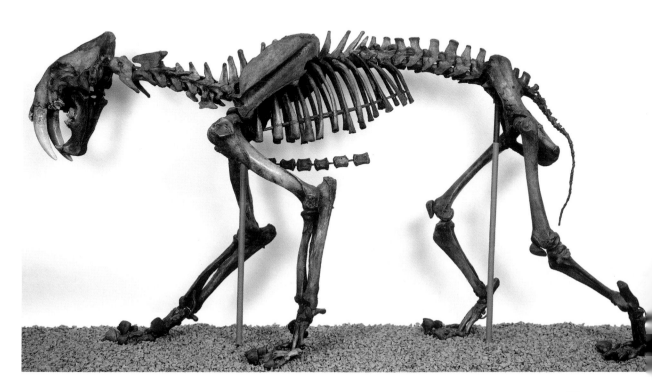

desperate animals might have mistaken those La Brea pools for water and come to drink. They would quickly realize their mistake but, by then, it would already be too late as they were caught in the thick, sticky tar, without hope of escape. A giant bison trapped in this way would appear an easy target for a group of predators but, as they closed in for the kill, they themselves would become trapped, and share their intended victim's fate. Birds uncovered here include

vultures, ensnared in the same way as they flew down to devour the dying, helpless animals.

Once they were caught by the tar, animals would have quickly sunk into it – so quickly that skeletons, and even on occasion pieces of skin and tufts of hair, were preserved. Rapid burial is one of the most effective means of ensuring preservation, and tar itself is a natural preservative – it can prevent the decay of even the smallest, most fragile bones of an animal's skeleton, such as those of the ear. This degree of preservation is bettered only by the permafrost of Siberia, which has preserved whole mammoth carcasses.

For years, the La Brea pits were known as the Salt Creek oil fields. Local people used the tar in building work, and, when they came across strange bones, originally thought they belonged to unlucky cattle that had wandered into the pits and were trapped. It was not until 1901 that the first scientific excavations took place. A century later, La Brea is world-famous for its fossilized remains – a total of about one million bones from a range of species have been gathered at the site, dating from 40,000 to 8,000 years ago. Clues about the prehistoric world are continually being unearthed, and palaeontologists such as Blaire van Valkenburgh are using what they have found to reconstruct the animals that roamed around the district of Los Angeles 10,000 years ago and more.

The bones uncovered are, as van Valkenburgh points out, a most amazing resource. Not only were the animals quickly buried – their bodies would have been intact to start with. They did not lie on the surface, to decompose in the normal way or be eaten or pulled apart by scavengers, nor were the bones scattered by natural disasters such as earthquakes. The fossil remains thus provide an invaluable aid to science in piecing together what life in this area was like at the end of the ice age. Palaeontologists have so far discovered no fewer than fifty-nine species of mammal and thirty-five species of bird. Many of the mammals were of the massive type already mentioned. A giant sloth's leg bone twice the height of a man has come to light, for example, not to mention the massive skulls of mammoths; and with the herbivores' bones lie those of the creatures that preyed on them.

From their fossil remains, it is possible to paint a pretty accurate picture of what a sabre-tooth looked like. Its most striking feature –

and the one that has captured people's imaginations – was its upper canine teeth, the ones that give it its name. These 'sabres', around 17 centimetres long but only 1.5 centimetres thick, stuck out beyond the jaw when the mouth was closed, giving the animal a fearsome appearance that would have been emphasized when the mouth opened. The bone sockets at their base show reinforcement, indicating that the teeth were used in a stabbing gesture. When they had done their job, it was the turn of the incisors to tear flesh from the carcass of the prey animal; the sabres were far enough behind the incisors to allow this. The thinness of the sabres indicates that they were in fact quite delicate, and they were not exceptionally sharp either, though they were serrated front and back towards the tip. They were curved, slender and knife-like, aptly named after the sword of the same shape.

Smilodon's overall facial appearance is generally accepted as being more like that of modern big cats than any other animal (although it is not their true ancestor). There are, however, different opinions about the precise shape of such features as ears and mouth. The American palaeontologist George Miller has suggested three major points of difference for the Smilodon's head, for example, which mark it apart from modern cats. Basing his argument on analysis of the skull, he maintains that the ears would have been set much lower on the head than a modern cat's, and that the nostrils would have been in a retracted position (the nasal bones are short), which would make the animal look somewhat pug-like, especially with the line of incisor teeth well to the fore of the canines. Miller also suggests that the lips would have had to be longer and more doglike in appearance in order to accommodate the animal's wide gape.

Certainly the Smilodon's body did not resemble a modern feline predator – its skeleton reveals a unique shape, quite unlike any cat alive today. It was about the size of a modern lion, but twice its weight, with a stockier body and shorter legs and a barrel chest; and it had a bob-tail. In fact, it looked more like a bear than a big cat. Its forelimbs were massive – twice as thick as those of a jaguar – and stronger than its hind legs. This was an ambusher, not a pursuer, and without doubt a highly adapted and extremely powerful predator.

In common with almost all members of the cat family, it had retractable claws, whose long, curved shape would have been ideal for spearing the body of a quarry and holding it firmly. As the zoologist Alan Turner pointed out in his book *The Big Cats and Their Fossil Relatives*:

> In prey capture, the extended claws can be dug into the skin of an animal and will enable the cat to cling on; if it chooses to, the cat can close the extended claws much as we would close our own fingers round an object, further digging the points into the skin of its quarry and making escape very difficult. This ability also serves to help the cat climb trees, but it can make descent somewhat hazardous because the deployment works better for ascent and provides little grip on the return journey. This explains why your domestic cat can virtually run up a tree but may have to be rescued with a ladder.

An artist's reconstruction of the sabre-toothed tiger's head.

It is unlikely, however, that sabre-tooths were climbers, on account of their great weight, though they may well have been capable of it.

As for its colour, its short fur may have been sandy, with darker stripes and patches providing camouflage; while there is no basis but supposition to support this idea, it seems reasonable. Most modern cats have some kind of facial marking, even those such as the puma that have plain bodies. Also, the young of some modern unmarked species, again including the puma, go through a stage when their coats are spotted, which may indicate a spotted coat in the animals' ancestry.

Given this physical picture of a formidable predator, the next question is just how the animal would have tracked and killed its prey. It is likely that sabre-tooths were daylight hunters, and that they did not have perfectly aligned stereoscopic vision. Machairodonts' skulls generally have small eye-sockets, indicating small eyes that probably did not adapt well for night-vision. This contrasts with modern cats, whose skulls, along with those of their ancestors, have large orbits, indicating large eyes, capable of seeing in the dark. Little is known about its hearing ability and, because cartilage does not fossilize, the shape of its ears can only be guessed (leading, as noted earlier, to George Miller's controversial conclusions). However, the sabre-tooth had a relatively advanced brain, showing some complexity in the areas controlling vision, hearing and limb co-ordination.

The sabre teeth were undoubtedly designed primarily for killing. (It is interesting, incidentally, that wear on the animal's canines seems to have been generally less than on its other teeth.) They could not have been used for biting into the prey while the prey was moving – there would have been too much risk of a tooth snapping off. The prey would first have to be immobilized, but, in order to do this, the tiger would have to get close enough to its prey to pounce on it. As noted earlier, its musculature was not that of a pursuer, a sprinter. All the power of the body was best suited to an ambush hunter, and its strength would have been concentrated in a final, lethal pounce. In fact, while modern big cats can and do run their prey down – lionesses and cheetahs are the clearest examples – many prefer the more energy-efficient ambush/pounce approach. By comparison with dogs and hyenas, cats are not well adapted for sustained running. The sabre-tooth was probably capable of a 'half-bound' – a short kind of gallop that would have used a lot of energy but, because of its quick acceleration, would have been perfectly adequate when needed to rush a quarry.

So the sabre-tooth would have used its weight, strong forelimbs and retractable claws to capture and knock down its prey. It would then have held it still and bitten into its throat or belly, able to cut through skin even as tough as that of a bison. The fact that the sabres were serrated meant that they could cut and slice and be withdrawn at a different angle from that at which they went in – something that the sharp but conical canines of modern big cats cannot do. It is not possible to demonstrate this experimentally by simply holding a fossil tooth and attempting to cut into bison hide; the action of the teeth had to be powered by exceptionally strong muscles, and opposed by a strong jaw.

There is no agreement on how the *Smilodon* actually dispatched its quarry, however. In the past, scientists have proposed that the animal drank its victims' blood, or that it used the knife-like teeth to stab its prey fatally, then withdrawing and allowing it to die before returning to feed. Some naturalists have put forward the idea that the teeth were primarily for sexual display, and that they were a hindrance to efficient hunting. However, since three other ancient groups of predators evolved sabre teeth – the nimravids, the creodonts and the marsupials – their effectiveness must be accepted. What their strike rate was like is unknown: cheetahs have been observed to capture about 70 per cent of quarry chased; lions manage only 35 per cent. But cats tend to take prey animals near to the upper limit of their ability, which makes sense, because the larger the prey in a ratio to the effort spent in catching it, the greater the reward.

In order to explore further the means by which the sabre-tooth killed its prey, the palaeontologist Virginia Naples has been working on the dissection of modern big cats to see if by comparison she can deduce what forces the sabre-tooth needed. By building up a better understanding of the animal's muscles, scientists can work out what it may have looked like, how it moved, and exactly what it was capable of in terms of strength and power.

It is not an easy job, since the machairodonts, the sub-family to which the sabre-tooths belonged, is not in a direct ancestral line to modern felines. However, the living cat that seems most closely to resemble it is the forest-dwelling jaguar. By comparing its skull with that of the sabre-tooth, Naples is able, by virtue of the known muscu-lature of the jaguar, to make certain deductions about that of the sabre-tooth. They reveal, for example, that the sabre-tooth had a very large temporalis muscle, one of the main muscles controlling the jaw, and that the sabre-tooth could open its mouth very wide – much more so than a jaguar can – and thus deliver a devastatingly strong bite. Its gape was much wider than modern big cats' – 90–95 degrees as opposed to 65–70 degrees.

The attack would have been helped if (as seems likely) the prey animal had gone into a state of traumatic shock, a kind of self-paral-ysis that would have provided the *Smilodon* with the static target it

A mould taken from a sabre-toothed tiger skull and lower jaw.

needed to sink its fangs into without risk of their breaking. The phenomenon of shock in the victim has been graphically described as its 'being a witness to its own execution'.

There is, however, some debate about whether the belly or the throat would have been the preferred target: one school of thought favours the belly, as it provides a greater area and furthermore it would be easier to hold an animal still and down on its back if the predator were biting into the stomach. On the other hand, the only known predator that favours this approach is the giant monitor lizard, the Komodo dragon; and a faster kill would be achieved by biting through the principal veins and arteries of the throat – which itself presents a more specific target. The belly would present a flattish surface, difficult to press sabre teeth into. This has been precisely demonstrated by Virginia Naples and her colleague Larry Martin, curator of the Museum of Kansas, in experiments with a metal replica of the jaw of a *Smilodon*. There is also the question of the relative difficulty of rolling a large prey animal on to its back to get teeth into its belly (while at the same time avoiding its kicks) in the first place. Martin and Naples think that sabre-tooths killed their prey standing up.

Smilodon had, as we have seen, short hind legs. It was also able, bear-like, to put the whole sole of its hind feet flat to the ground, giving it a solid base to support the push-and-pull effort of wrestling with the upper quarters. From fossil remains found at La Brea, it is evident that sabre-tooths had lumbar problems, just as humans do, pointing to a frequently held upright stance. Martin's and Naples's idea is that the killing method was similar to that of a bear. To control a big prey animal, the sabre-tooth would have had to grasp its head and pull its neck round, thus exposing and tensing the neck so that it could be bitten into, a procedure which if correctly followed would result in more or less instant death. The fact that the sabre-tooth was bear-like in build supports this theory.

The ambush method of attack presupposes the right kind of terrain in which to hunt. A likely place for a kill would have been a waterhole or lakeside to which prey animals would come to drink, so that nearby trees, high grasses or shrubs would have been essential cover for the sabre-tooth to use. Analysis of pollen from the late ice

SABRE-TOOTHED TIGER: VITAL STATISTICS

Smilodon fatalis

Appearance: possibly tawny with some camouflage markings on flanks and face. Powerfully built with massive shoulders and forequarters. Distinguished by long, blade-like canine teeth.

Size and weight: about that of a modern African lion, but twice its weight.

Habitat: grassy plains and open woodland.

Longevity: not known. Perhaps nine years.

Distribution: specifically in North America, though related species worldwide in earlier times, and a related species was contemporary in South America.

Reproduction: perhaps up to four cubs annually after a gestation period of 4–5 months; young at least semi-independent for up to first two years and female young may have associated with mother after maturity. Young at least semi-independent up to end of first year of life.

age suggests that such vegetation would have existed. The habitat of the sabre-tooth would have been a combination of tall grasses with deciduous forest. However, even with the element of surprise, sudden attacks were not without danger for the sabre-tooth, since the prey animals were large and powerful and well able to defend themselves. The fossil record reveals many skeletons in which bones have been smashed – even in its death-throes a bison could deliver a last, powerful kick.

The sabre-tooth's preferred prey would almost certainly have been bison, and the wild horses. It may have taken young mammoths, too – probably at the slightly advanced age when curiosity led them to wander just far enough away from their mothers to allow an attack.

Palaeopathologist Chris Shaw has examined sabre-tooth skeletons that reveal the kind of injuries the animals sustained: gashes in a skull near the eye socket are evidence of fierce fighting. Sabre-tooths fought among themselves on occasion, possibly over kills. In 1932 palaeontologists from Nebraska unearthed a fang from an archaic species of sabre-tooth that still skewered the shoulder-blade of another sabre-tooth. The 25-million-year-old remains bear testimony to an ancient catfight that was probably fatal to both participants. Other injuries identified more recently by Shaw include dislocated hips, bite wounds, and perforations along the spine.

The most common injuries, however, were to the front limbs and the lower back, and these injuries are such as to suggest that the animal sustained them while hunting. As Shaw points out, 'Bringing down a ten-ton bison would often result in injury. The price of survival was high – we've found 5,000 sabre-tooth cat bones that are a mangled mess, and these are the ones that actually survived.' Examination of fossil bones often shows that there was secondary growth of bone in the region of a muscle attachment site. The insertion areas of the deltoid muscle on the humerus is specially thus affected, probably as the result of stress during sideways movements of the limbs of a kind that would occur when attempting to subdue a large, struggling quarry.

The sabre-tooth's hunting methods lead to a consideration of its sociability. Most living cats (except lions, which live within a highly structured social set-up) are solitary creatures, coming together only for mating. But probably it would have taken more than one sabre-tooth to bring down a giant camel or deer – which seems to suggest that sabre-tooths were sociable animals. Opinion, however, is sharply divided. Collaborative hunting is generally used by open-country hunters, such as lions and cheetahs. Jaguars, leopards and tigers operate in wooded terrain, and there the element of ambush is more important; collaborators in such circumstances may be more of a hindrance than a help. Sabre-tooths hunted in a mixed terrain, but so do pumas, and they are solitary creatures.

It may be that sociability is controlled by other factors. That most solitary of animals, the snow leopard, will live amicably with its fellows in a zoo, for example, as will domestic cats in the home – and they, despite spats, often seem happier as pets if there are more than one of them. Among jaguars, leopards and tigers, each male has a territory which is firmly boundary-marked for other males, but which intersects the smaller territories of several females. It is now thought, through observation of tigers and leopards, that a given male will be far more tolerant of both females and its own cubs within its own domain than was previously believed. Alan Turner writes: 'Family groupings have even been seen, and in some cases leopards have been observed to mate while the female still has cubs. There is also now some evidence that, as with lions, it is the juvenile leopard males that are likely to be less tolerated by the mother and forced to disperse, while the female cubs may remain in the vicinity for longer and even establish territories adjacent to hers.'

While hunting, a sabre-tooth would leave scent marks on trees or shrubs as it passed – not only as a means of marking a boundary and securing a territory, but also as a means of exchanging information with others of its kind. Cats are capable, too, of a variety of facial expressions and vocalizations, all of which suggest sociability, since otherwise there would be no need for such skills. And while roaring is a sound designed to carry over distance, and may in some contexts serve as a 'warning-off' device, purring is a quiet and intimate noise, entirely suggestive of social acquiescence. (Analysis of preserved sabre-tooth skeletons has revealed the existence of the hyoid bones in the throat – showing that the animal was capable of roaring in much the same way as a lion.)

Of course it does not necessarily follow that all this information can be extrapolated and applied to the sabre-tooth, but the arguments for at least a degree of sociability seem convincing. One counter argument cites the fact that the brain-size of *Smilodon* is similar to that of the jaguar and the leopard; this, together with the high number of lesions on bones of the sabre-toothed tiger within a locality, is taken to indicate low levels of sociability and high levels of intraspecific fighting. But from the new evidence regarding leopard behaviour, this view may have little foundation.

Maternal care was almost certainly important. From analysis of their dentition, it is likely that the animals did not reach full maturity and independence early – the famous sabre teeth were the last to develop, not until well into their second year. But they would have been strong and big enough well before that time to be able to assist their mother in bringing down a large prey animal – and in the process would have learned hunting techniques.

Furthermore, the fossil record suggests that wounded animals of the species were looked after by their fellows, as happens with lions now. Some damaged bones show signs of successful healing, which indicates that their owners lived for some time after sustaining the injuries; and many others seem to be the bones of aged or otherwise impaired individuals, whose survival is in itself an indication of sociability. Alan Turner adds:

> *A further argument against the solitary and aggressive lifestyle of* Smilodon *has been put forward based on the sheer numbers of individuals recovered at Rancho La Brea and a few simplifying assumptions. First, it seems likely that the cats became mired in attempts to reach already-trapped animals or their carcasses. If the number of large herbivores found gives an approximation to the maximum number of trapping incidents (a plausible suggestion) then several individuals of* Smilodon, *perhaps ten or so, were also trapped during their efforts to obtain each of those herbivores. Since it is unlikely that every* Smilodon *attempting to reach a trapped herbivore did indeed suffer the same fate, the implied number of animals in the vicinity at any one time, even if the trapping incidents were themselves infrequent, is really too high to*

support an argument for a solitary lifestyle, since such a lifestyle would imply discrete territories and would reduce the number of animals able to congregate in the area.

At the La Brea tar pits, it has been estimated that for each herbivore found there were nine or ten predators, of which number 2.8 on average were sabre-tooths. While the remains of over 2,000 sabre-tooths have been found at La Brea, a total of only 700 horses, bison, camels and so on have been extracted. The ratio certainly suggests that the predators hunted co-operatively.

On the other hand, do the La Brea pits give a false impression? Given the system of the food chain, it is accepted that any one animal type will be fewer in number than its food source – so there are more herbivores than carnivores. Generally speaking, the fossilized remains of cats should be fewer than those of herbivores; but thanks to the fact that they were cave-dwellers, and thanks to such phenomena as tar pits, more than their 'fair share' of remains were available to be discovered.

Interestingly, no evidence has yet arisen to suggest that there was any great size difference between the sexes of *Smilodon*. This fact supports the view that it is unlikely that sabre-tooths formed lion-like prides. Lions are much larger than lionesses – a fact that is pivotal to the social organization of a pride.

But another point against the solitary *Smilodon* is that the sociability of cats depends on their environment, and the density of their prey. If an area of land supports fifteen herbivores which in turn provide prey for one sabre-tooth, then the picture is clear. But let us say that the same area supports forty-five herbivores. Logically, they would feed three sabre-tooths independently, but if they were not evenly distributed in their area, but moved around it in a herd, then it would also make sense for the three sabre-tooths to pool their resources and hunt in a group. As the ice age came to an end, there was a large number of big herbivores, which theoretically at least suggests that a significant group-hunting population of predators existed alongside it.

Furthermore, a small group would be better able to defend a kill from other predators, such as hyenas or vultures, than a solitary animal. An adult sabre-tooth would have needed about 30 kilograms of meat a week to sustain itself, so a good large kill would represent enough food for a week for several tigers – more than enough for a small family group of three or four. It is thought likely that the animal was able to gorge, after which it would spend some time digesting, thus obviating the need to make a kill all that frequently, perhaps only once or twice a week – thus compensating for the energy expended in bringing down a really large quarry.

While such a system continued, life would have been good for sabre-toothed tigers. They had no enemies, and no real direct competition, except perhaps from scavengers and opportunists, such as hyenas and dire-wolves. (The only problem the sabre-tooth would have had, given its 'delicate' teeth, would have been in hauling an uneaten carcass to a place safe from scavengers, or, if solitary, in defending an uneaten carcass from other predators.) So *Smilodon* was ideally suited to its environment – as long as the large prey to which it was specifically adapted continued to exist in sufficient numbers. But, as was so often the case, its over-specialization meant that when sudden change came, it was not able to adapt; and towards the end of the ice age, such a change was on the way.

The fossil record enshrined at La Brea offers a wealth of clues and information about how animals of the late ice age lived, but very few concerning how they died out – a process that in the majority of the large species then alive occurred very rapidly. The palaeoecologist Greg MacDonald has noted that about 11,000 years ago sabre-tooth remains at La Brea peter out. Examining the stratification in the pits, above the 11,000-year level, he says, 'We stop finding sabre-tooth bones. There isn't even one.' The last skull found at La Brea dates from 11,100 years ago, though a related species seems to have lasted another 2,000 years in the area now occupied by the town of Nashville. 'This is a species that has been around for a million years and suddenly there's a mass extinction,' says MacDonald, 'and at the same time the ice age is coming to an end. A new era is dawning, which is the one we are now in – the Holocene.'

But how did such a successful animal disappear quite so fast? One answer may lie in the dramatic changes to vegetation brought by the warmer climate that heralded the end of the ice age.

Palaeobotanist Julio Betancourt has been working in Arizona, investigating evidence unwittingly left behind by a small, modest mammal that existed at the time of the sabre-tooth, but which survived the changes and is still alive today. The packrat is of great value to scientists working in the field of palaeontology because of its habit of building middens composed of local vegetation, which becomes glued together and preserved by the action of the packrat's urine. Ancient middens dating from the time of the sabre-tooth and earlier have been preserved in caves and in them, equally well preserved, can be found the seeds, leaves and fruits of plants that lived anything up to tens of thousands of years ago. By carbon-dating them, scientists can discover not only which plants grew locally, but when; and by locating middens dating from just before and just after the end of the ice age, they can see what happened to the plants.

The first step is to place the mass of vegetation in water for a week, a process that dissolves the urine and allows the mass to separate into its component parts. Microscopic examination of those parts reveals how the nature of the vegetation changed. What Betancourt and his colleagues have discovered by comparing many different middens from the same area is that there were very big changes in vegetation between 10,000 and 12,000 years ago. The Kofa mountains of south-western Arizona, only about 500 kilometres west of Los Angeles, now stand in desert, but 12,000 years ago the region consisted of open woodland, and this is a change that has been observed over large parts of the south-western USA. The Los Angeles Basin became warmer and drier at the end of the ice age. River beds dried up, and the trees that once lined the banks consequently disappeared.

Concentrating on the evidence provided by the La Brea tar pits, it appears that around 40 per cent of the animal species found there

went extinct at about the same time – about 11,000 years ago. The other 60 per cent survived. Why should this have been so? Greg MacDonald believes that as far as herbivores are concerned, the split was between single-stomached animals, such as horses, and ruminants – multi-stomached creatures, such as bison. The ruminants survived, and indeed bison thrived. The indigenous horse, the Columbian mammoth and the giant sloth, to give three examples, did not. It may be that a ruminant's longer digestive process was better able to cope with changes in vegetation – as evidenced by a bison's dung, which is smoother and more broken down than a horse's.

But why, then, did the sabre-tooth also disappear? If the horse had gone, then the bison survived as a prey species. Why could the sabre-tooth not hunt the surviving herbivores?

One key factor was that the ruminants themselves started to change. The new generations were smaller, faster animals that moved around in larger herds than before. Larger herds meant better defence, and it was harder for a sabre-tooth to single out one animal. As the landscape became more prairie-like, and trees and shrubs decreased in number, so did the opportunity for ambush, since there was no cover left, and the sabre-tooth was not designed for pursuit. And as for feeding on smaller prey than it had been used to, there was another problem. The vast majority of species that survived the transition from the ice age to the early Holocene were too small for the sabre-tooth to eat. It was designed to hunt big game only, and its whole adaptation reflected that specialization. Small deer and pronghorn were too alert and too fast for a sabre-tooth to have a hope of catching.

The very qualities that made the sabre-tooth such an adept hunter of giant herbivores now acted against it. What is more, the sabre-tooth was not well adapted for scavenging, and in any case would have found it hard to compete with other large carnivores such as bears in this area. The numbers of the animal began to thin out, reproduction became more difficult, and the way to extinction was paved.

At about the same time – though some scientists now believe that his arrival was much earlier – man was making his first appearance in North America. As we saw in the previous chapter, Clovis Man crossed the land bridge between Siberia and Alaska over what is now the Bering Strait, swiftly sweeping southwards. These nomadic hunter–gatherers would, it is believed, have depopulated an area of game before moving on; as their throwing spears were equipped with the highly efficient flintstone Clovis Point, they were deadly killers and would have been dangerous competitors for what prey food was still available to the dwindling sabre-tooth population.

Man may have established himself throughout North America by 10,000 years ago; if he did come into contact with the sabre-tooth, it would have been his dominance in the battle for food rather than through direct slaughter that he would have edged an already doomed species over the brink of extinction more quickly. Species that went extinct in the last ice age had survived others: in the relatively recent past, ice ages have been a regular occurrence, cropping up at 100,000-year intervals over the past 800,000 years. The exceptions have been the interglacial periods – each of some 10,000 years (today, we are reaching the end of such a period). So some scientists believe that in the last ice age, man represents the one new factor that clinched the fate of endangered species.

But there is no agreement. Greg MacDonald maintains that there is no evidence that the sabre-tooth was ever hunted by man, or indeed that man was anywhere near La Brea until well after the sabre-toothed cat had died out. Although it's possible that man was responsible for wiping out the sabre-tooth's prey, hunting mammoths and bison, the mammoths died out and the bison survived. MacDonald is sceptical that a small population of hunters could have been responsible for such a mass extinction.

Blaire van Valkenburgh on the other hand holds that the huge change in climate and vegetation, which disrupted the ecosystem, was a key factor, and human hunters may well have delivered the final blow, attacking the stressed animals and pushing them over the edge, killing their prey and eliminating their chances of survival.

There is, however, no evidence as yet of man coexisting with the sabre-tooth. And although the sabre-toothed cats hung on in North and South America until about 10,000 years ago, they had gone extinct in Africa 1.5 million years ago. Extinction is an element of life's progress: nearly all the flora and fauna that have ever lived are now dead. But in this particular case it is interesting to ask why the sabre-tooths were able to hang on so much longer in the Americas.

There were considerable differences in climate: Africa bestrides the equator while North America lies above it. Also, prey animals on the different continents comprised different species. The Americas did not have antelopes until considerably later than Africa, which always had a wider choice of prey. Climatic changes in Africa about 3.2 million years ago affected the herbivore populations; lighter, faster animals developed, and this in turn led to an answering change in the predators.

The debate continues, but it will never be possible to attribute a single cause to the demise of the sabre-tooth. It was an animal which, like so many at that period in time, was overtaken by events and changes in the world that happened so quickly that it did not have time to adapt to them. Their niche in the world had disappeared, and it was time for them to go.

ANIMATRONICS

Not only did the animals in *Extinct* have to move, they had to appear to do so within the real world, interacting with living creatures and humans. This is very hard to achieve convincingly. The solution we chose was to build high quality animatronic models.

Animatronics, as the name suggests, are moving robotic models. Since they're actually real, rather than just data in a computer memory, they can be filmed interacting with the real world. They can be placed in real locations, they can eat real food, and they can interact with actors or even real animals.

They can also be filmed in close-up, which is much harder to do using computer-generated imagery (CGI). Computer graphics allow us to make a wildlife documentary about an animal that no longer exists, but as well as wide shots of the whole creature, or group of creatures, we want to see close-ups – the head, the eyes, the mouth and teeth. While it is possible to do that using CGI, the degree of detail required is vastly increased, and it's easy to see why. Imagine the complexity of a tiger's eye – how the light shines in it, how it moves and flickers. Take the wetness of the tongue, and the complex array of movements it makes while the beast is eating. Or look at the hair on a human head and imagine a close-up of the mass of different shapes all forming the head of hair we perceive as a single entity from far away.

The interior mechanism of the Tasmanian tiger head.

Once the decision to commission animatronic models was made, the project was passed on to Artem, one of the UK's leading movie model and special effects companies, with a long track record in the industry and a reputation for technical know-how and creative ingenuity. Work began on building accurate heads of the

Building the Tasmanian tiger head: latex mould (above); skin surface (left); finished model (below).

Foam mould of the dodo head.

sabre-toothed tiger, great auk, Tasmanian tiger and dodo. In addition, we decided we'd need a mammoth's trunk that would appear to feed and be able to interact with its environment. The process of building the animatronics began well in advance of the computer animation, because they needed to be ready for the location shoots. Models accurate enough to convince people they were seeing the real thing naturally require a great deal of work.

The first stage was painstaking research. The basic shape of the model would be sculpted in clay, so the sculptors had to be absolutely familiar with the animals they were re-creating. Liaising with production zoological researcher Kate Dart, the team visited museums, taking photographs of fossil bones, stuffed birds and reconstructed models. Scientists were consulted to get their views on details that fossils simply can't provide, like eye colour and shape, and the texture of fur and feathers. Even where actual stuffed examples exist, nothing could be taken at face value. Stuffed great auks and dodos are rare, and they're also very old, dating back to when the creatures became extinct. Time has taken its toll on these curiosities – sunlight has faded the colours, and the soft tissue of the eyes and tongue have crumbled away. Even the physical shape of the animal, as rendered by the original taxidermist, might be inaccurate due to their own limited knowledge of the animal in the wild.

The Artem workshop.

When sculpting was complete, it was amazing to see just how much personality the creatures took on. Standing next to a life-size sabre-toothed tiger head, the sheer physical power of the beast becomes apparent. The dodo, so often maligned in history as weak and stupid, is actually imposing and even fierce. And the Tasmanian tiger, far from being goofy and kangaroo-like, takes on an altogether more menacing appearance. And all this before they had even moved! The clay heads and mammoth trunk were then covered in latex and moulded to make rubber models ready to have more distinct features carved into them. The most time-consuming process of all was furring, feathering and colouring the resulting models. Adding fur or feathers to the latex is usually done by hand, and several fur types have to be carefully applied to a single head in order to make it look convincing.

How did we then go about moving the models? To figure out the best method requires an understanding of how they would be shot and used in the finished film. Only using models for close-ups meant that Artem's operators could manipulate them on location, with no need for complex wireless remote controls. Instead, each model was operated by hand – like an expensive glove puppet, effectively. This low-tech approach has real advantages, since the movement generated by a human hand is far more accurate and convincing than all but the most advanced robotics. Built in to the models are the complex mechanics necessary to make the animatronics move. Micro-engineering was required to make flaring nostrils, rolling eyes, dilating pupils and licking tongues, and these tiny motors were remote-operated by a crew member in situ while another played the 'character' of the creature. To add extra realism, liquids that mimic saliva, blood, and nose and eye moisture were applied to the models immediately prior to shooting.

Once filming was under way, the process of bringing these long-dead creatures to life really began.

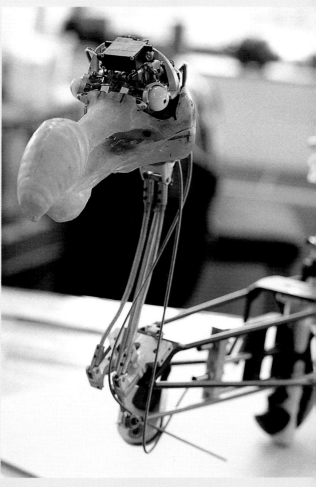

Interior mechanism of the dodo head, including radio controlled, moving eyes.

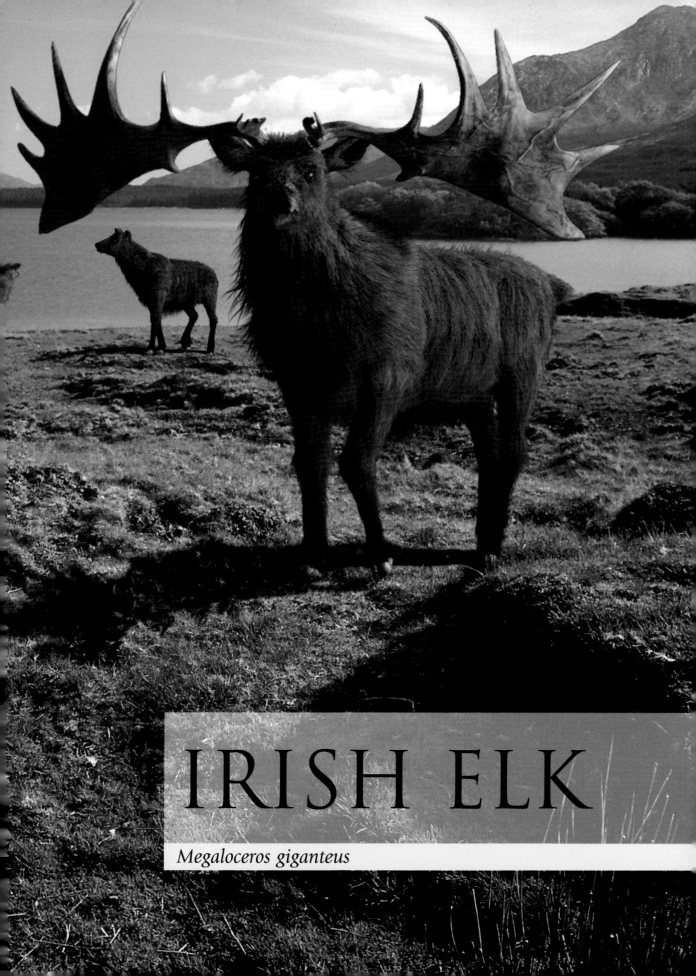

IRISH ELK

Megaloceros giganteus

Ten thousand years ago, the last great ice age was coming to an end. Ireland then had not yet been colonized by man. Its rolling hills were covered with grasses and scattered shrubs and trees. There was little snow, and the climate was gradually becoming more temperate, but it was still a forbidding place. Nevertheless, many creatures lived and flourished there, among which the undisputed monarch was the Irish elk.

It was more spectacular than any animal that had roamed Ireland before, or would since. It stood nearly 2 metres high at the shoulder, was 3 metres long, and weighed up to 700 kilograms. But

A nineteenth-century French engraving of the Megaloceros skeleton.

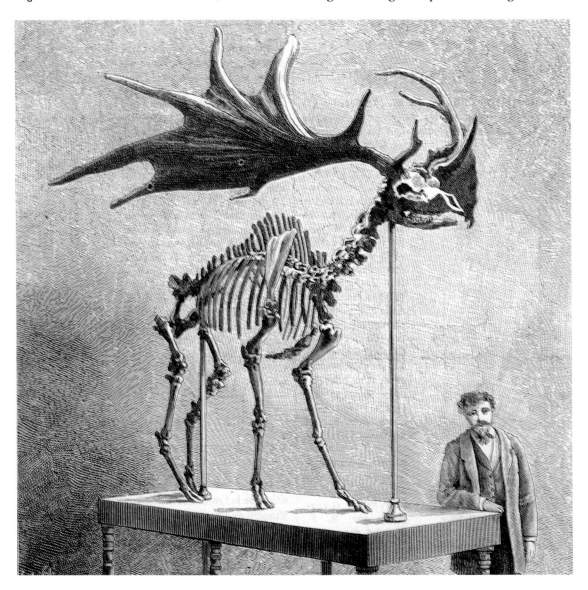

what really made it sensational were its antlers. They were elaborate to a degree that can only be called baroque, weighing 35 kilograms a pair and spanning up to 3.5 metres. They were outspread at right angles to the head and broadened into a palm that faced slightly backwards and upwards when the head was held horizontally. (Only modern fallow deer have a similar 'palmation'.) The largest set known is to be found today in Adare Manor, the home of the Earl of Dunraven, in County Limerick. 'Size has a fascination of its own,' as the great biologist Sir Julian Sorell Huxley once wrote.

But something over 10,000 years ago this regal creature, which had survived and flourished throughout the ice age, stood on the brink of extinction – which, when it came, did so with astonishing speed.

Many fossil remains have been uncovered, especially in Ireland, which has the largest concentration anywhere. The animal's domain spread throughout Europe, almost as far south as the southern coast of Spain, as far north as southern Scandinavia, and far into central Asia, almost to Mongolia. Remains have also been found in Austria, Britain, France, Germany, Hungary, Northern Italy and the Middle East, and a related species has been found in China. Smaller and simpler primitive forms of the animal are found in the fossil record in Europe in the Middle Pleistocene period, and the first true Irish elk appeared circa 400,000 years ago, and was still comparatively small. The ancestry of the Irish elk has not yet been fully determined, however. It has played a role in myth and legend, and there is evidence that early man encountered it. Once it was (mistakenly) identified with the *schelch* of the *Nibelungenlied*. The origin of the word 'elk', however, derives from the Old English *elh*, the Old High German *elaho*, and the Old Norse *elgr*, among other roots.

Dr Thomas Molyneux wrote the first scientific essay on the Irish elk in 1697, mentioning a fine skull, with antlers attached, found near Dardistown in Drogheda: 'We have a remarkable example in Ireland, in a most large and stately beast, that undoubtedly has been frequent in this kingdom, tho' now 'tis clearly extinct; and that so many ages past, as there remains amongst us not the least record in writing, or any manner of tradition, that makes so much as mention of its name...'

He was, however, under the impression that although it had died out in Ireland, it must still be alive somewhere, and his was a commonly held belief. During the seventeenth and eighteenth centuries, it became increasingly clear that many fossils discovered were of creatures that no longer existed, at least in the known world, and certainly not in the places they were found. But this presented a deep conflict in the minds of early scientists who believed unquestioningly in divine creation. Any idea that God would allow any of his creations to disappear would have been heretical – it would have been to think the unthinkable. The reality of extinction was something even the most advanced thinkers were slow to accept.

Since general faith in the ultimate bounty of God precluded any sense that any beast should have ceased to exist, Molyneux was echoing an idea held by nearly everybody at the time when he wrote: 'That no real species of living creatures is so utterly extinct, as to be lost entirely out of the world, since it was first created, is the opinion of many naturalists; and 'tis grounded on so good a principle of providence taking care in general of all its animal productions, that it deserves our assent.' He thought that the North American moose, then little-known but of which sketchy accounts had crossed the Atlantic, was probably the same animal as the one whose bones had begun to be discovered in bog- and peatland almost all over Ireland. He put the reason for its local disappearance down to an 'epidemick distemper' brought on by 'a certain ill constitution of air'.

It was not until after the investigations of the seminal French naturalist Baron Georges Cuvier, early in the nineteenth century, that the Irish elk was deemed to have been a victim of the last 'great global catastrophe', the ice age. This dictum, which embraced not only the Irish elk but other fossil vertebrates such as the mammoth, was a crucial step forward: it urged the acceptance that animals once walked the Earth which were no longer here at all. In 1812 Cuvier declared the Irish elk to be *'le plus célèbre de tous les ruminants fossiles'*. After him, various theories were advanced to explain the animal's demise, from a deluge to the depredations of Celtic tribes or the Roman amphitheatre.

Since its discovery in the late sixteenth century, the Irish elk has been an object of intrigue for science and, on account of its 'kingly crop', a magnet for collectors and trophy-hunters. In Ireland its antlers have served not only as decorations for baronial halls, but as temporary bridges over streams and adornments for gateposts. Yet the circumstances of its disappearance remain mysterious: why had it been so successful during the harsh conditions of the ice age, only to disappear suddenly as conditions improved?

The Irish elk was neither an elk, nor exclusively Irish. In North America the 'elk' is the animal called by Native Americans the wapiti; in Europe, 'elk' is used to describe the moose. The Irish 'elk' was related to neither creature. It should perhaps more properly be called the giant Irish deer, or even – given that 'Irish' was just a reflection of the number of fossils found in that country – simply 'giant deer'.

Megaloceros giganteus, to give it its scientific name, was not the biggest deer that ever lived: that honour goes to a North American animal, a deer known as *Cervalces*, which flourished between 50,000 and 10,000 years ago. Our 'elk' was about the same height as a modern moose, and perhaps on average a little lower at the shoulder. And although it had the largest antlers of any deer ever known, they were not disproportionate to its body size. Tom Hayden, Professor of Zoology at University College Dublin, has calculated that, in fact, the antlers of *Megaloceros* are exactly as large as the comparison would predict. Hayden looked at the relationship between body size (shoulder height) and antler size in the thirty-five species of deer living today – a small deer, like the Sika stag, which weighs about 50 kilograms, has antlers measuring about 40–45 centimetres long, while a fallow buck weighing about 110 kilograms has antlers about 60–70 centimetres long. By the same calculation, the *Megaloceros* result fits the pattern.

Professor Hayden is also the founder member of the Irish Deer Society, and has access to the store-rooms of the Dublin Natural History Museum. In an innocuous-looking warehouse on an industrial estate on the outskirts of the city, there is a treasure-trove containing thousands of fossil remains of the Irish elk. This store represents the fruit of excavations that have been taking place since the beginning of the nineteenth century, retrieved from peat bogs the length and breadth of the country. The Natural History Museum itself, in Merrion Square, has three huge black mounted skeletons of the animal, two males and a female, in its entrance hall. They are so large that some visitors have taken them to be a kind of dinosaur.

The earliest specimens worldwide were unearthed in England and Germany and date to about 350,000 years ago. Most of the Irish remains date from a short period leading up to the time of the elk's disappearance – between about 12,000 and 10,600 years ago. They have been chiefly found in lake sediments in shallow water that later became overgrown by peat, and are widespread in Ireland except in the extreme south-west. The most famous site of all, where over one hundred skulls were found, is Ballybetagh Bog, near Glencullen, about 15 kilometres south-east of Dublin.

Another, smaller, group of remains have been found in caves, and are much earlier: they date from between 40,000 and 31,000 years ago (the oldest remains discovered so far were at Castlepook Cave, 30 kilometres north of Cork). At this time, there would have been a land bridge to Ireland, which mammals could cross. This happened only rarely during the Pleistocene period; the Irish Sea is very deep, so water levels would have to drop dramatically before a land bridge could be formed. From about 31,000 to 12,000 years ago, Ireland would largely have been in the grip of ice at the height of the ice age, and it would have been impossible for the Irish elk to survive in such conditions. As the weather improved and the ice gave way to grassland, it probably recolonized its old territory from the south, where it had retreated along with many other species in the face of the encroaching ice, or came from Britain via Scotland or the Isle of Man.

The reason for the concentration of 'elk' fossils of late date in Ireland is probably due to the abundance of lakes. Frozen over through the winter, animals fell through the ice in the spring thaw, sinking into the mud at the bottom, where their bones were preserved.

The first record in Ireland of the elk dating from historical times comes in the form of a sketch of a skull unearthed in County Meath in 1588. The skull was displayed for a while at Rathfarnham Castle, and later found its way to Hatfield House in Hertfordshire, some time after its completion in 1611.

The giant deer, long an icon of Irish prehistory, was first identified as an extinct species only in the nineteenth century, and it became a scientific cause célèbre overnight. Clearly it had been a magnificent, open-country animal, having a big heart and stout legs, which made it an endurance athlete designed to outrun its only likely predator, the wolf, over long distances. It would have roamed prehistoric Ireland in herds of perhaps twenty-five or thirty, communicating with a variety of barking sounds. It had good eyesight and hearing, but its keenest sense was smell.

The enduring mystery of the beast, and the source of its fascination, lies in the antlers. A stag grows and sheds its antlers every year, an action that represents a great expenditure of energy. To grow antlers the size of an Irish elk's, a huge amount of energy would be required. What was the reason for them? The search for an answer to this question threw Victorian science into confusion, and appeared to give those opposed to Darwin's theories a chance to refute them. Nobody could explain what practical use such massive antlers could have, and it seemed illogical, the argument ran, to say that any creature would evolve in such a way as to produce so useless a burden.

A theory of evolution current in the nineteenth and early twentieth century, now abandoned, was orthogenesis, which argued that any change in a given organism was not due to natural selection, but to an internal and irreversible trend in direction within a species. Thus in the case of the Irish elk it was believed that the animal was set on an unchangeable course towards larger and larger antlers, which ultimately became a hindrance and brought about its demise. While there is a grain of truth in this, it presents too simplistic a picture.

The critics of Darwin took as their starting point this supposition that the Irish elk was the last in an evolutionary line of animals that grew ever bigger antlers, which became harmful to their existence, and were therefore part of a process that flew in the face of the idea of natural selection. The antlers' contribution to their owners' extinction prompted a fresh crop of theories, one of which proposed that a rush to the head from the antlers of the blood which had nourished them during growth brought on 'epilepsy or apoplexy'. Estimations of antler size became wildly exaggerated – one account gives a spread of 20 feet (nearly 7 metres). Even quite recently theories have been advanced that only the oldest stags would have grown antlers so big as to cause them survival difficulties – theories that overlook the fact that antlers are shed and regrown each year. In fact, while stags may grow heavier (and fatter) with age, the size of antler they grow annually will tend to decrease once they are past their prime.

IRISH ELK:
VITAL STATISTICS

Megaloceros giganteus

Appearance: one of the largest species of deer that ever lived. Possibly pale brown or dark yellow in colour, with a dark stripe running diagonally down flank from shoulder to hind legs. Large eyes and ears, broad, blunt muzzle for close-cropping of grass, its principal food.

Size and weight: nearly 2 metres at shoulder, and 2 metres long. Overall weight up to 700 kg. Outstanding feature was its antlers, the largest of any known deer, which weighed 35–40 kg the pair and spanned up to 3.5 metres.

Longevity: males are reckoned to have lived up to 13 years, females up to 16 years.

Distribution: throughout Europe and into central Asia.

Reproduction: mating season in August. Males and females separated in winter. Young at least semi-independent up to end of first year of life.

It is interesting that so many early naturalists assumed that huge antlers had to be disadvantageous. One reason for this seems to be that the Victorians thought that the antlers' only function was as weapons, either for combat with rivals or as a defence against predators. Darwin was among the few who realized that broad palms could hardly make antlers more efficient as weapons – rather the opposite. He toyed with the idea that 'they may serve in part as ornaments', going on to say that 'If then, the horns, like the splendid accoutrements of the knights of old, add to the noble appearance of stags and antelopes, they may have been partly modified for this purpose.' However, he does not pursue this line of thought: 'With mammals

A Megaloceros skeleton in the Dublin Natural History Museum.

the male appears to win the female much more through the law of battle than through the display of his charms.'

But if this were true of the Irish elk, then its rococo headgear had gone well beyond the call of battle; furthermore, even the moose, far more modestly equipped, has no need of more than its own body size, if it is in good condition, to protect itself from wolves. Apart from Arctic foxes, which would not have posed a threat, and bears, there were no predators other than wolves in Ireland at the time. And the Irish elk was not only massive, but also a powerful runner, totally adapted to its open environment – a landscape that would have provided little cover for wolves to mount any kind of ambush. But little exists in nature which does not have a practical application – hence the fascination of solving the problem of function.

The antlers' function continues to be the subject of debate, and even of controversy; but their role in the ultimate demise of the species remains a mystery, though, in grappling with the problem, naturalists are coming much closer to an explanation.

A reconstruction of a complete skeleton demonstrates how the stags had adapted specially to support the burden of their huge antlers. They must have had thick tendons connecting the back of the skull to the wide neck vertebrae, which in turn would have given rise to a pronounced hump on the animal's upper back. A stag in its prime, about nine years old, would have put the Monarch of the Glen to shame. To build up and maintain such a splendid physique, the Irish elk needed to be a grazer – an efficient bulk grass feeder – and evolved over many millennia into a creature capable of thriving on the mineral-rich grasslands of the ice age landscape. Ireland then would have been heathland dotted with low scrub – sorrel, willow, juniper, crowberry and sagebrush – and a few trees: willow again, and cottonwood, with some stands of birch. The Irish elk's successful adaptation to its environment would help account for its large size, though this, and the size of the antlers, also had a role to play in the sexual selection process of the species.

Analysis of fossil teeth under the microscope reveals the telltale silica scratches on the enamel which typify the teeth of an animal that feeds chiefly on grass. The same marks are found on bison, cattle – and the fallow deer, which many scientists (though by no means

all) believe to be the Irish elk's closest living relative. Another indication of the Irish elk's feeding habits is that it had an extremely broad snout compared to many species of deer. This shape of snout enables its owner to bite off the grass stems as closely as possible to the ground, and cram as much as possible into the mouth for chewing.

Grass alone, however, would not provide nearly enough of the calcium required to produce a crop of 35-kilogram antlers once a year. The Irish elk needed to seek out a more varied diet. It may, for example, like the moose, have eaten lakeside plants. Taking still-extant samples of food from between the teeth of fossil skulls, Tom Hayden has been able to identify that, among other vegetation, the stags consumed willow. Willow is rich in calcium and phosphorus, the bone-building nutrients they would have needed to grow their massive antlers.

Although this breakthrough in knowledge was a big step forward in unravelling the mystery, something else was discovered when the Irish elk's jaw was examined in greater detail. For a species of deer, the bone of the lower jaw was found to be very much thicker than normal. Scientists now believe that the density of the bone in this area would fluctuate in the course of a year. As winter passed, the willow shoots began to sprout and the stags would eat them. The calcium sucked out by the stags would accumulate in the jaw, making it very dense. The resultant fund of calcium could then be called upon during the late spring and early summer months as the stags grew their new antlers, ready for the mating season, which it is believed took place in August.

If the growing period for the antlers is assumed to start in April – and four months to achieve full growth is a conservative estimate by comparison with modern deer – nevertheless even a modestly sized set of antlers, of 2.6-metre spread, would have had to grow at a rate of 8.7 millimetres a day, as Anthony D. Barnosky of the University of California at Berkeley has pointed out. The energy this would have required is immense, and food consumption to sustain it must have been prodigious. It has been calculated that each stag would require 56 hectares of shrub-dominated land to meet its phosphorus requirement during the summer months, and a greater area if shrubs were not plentiful. Each stag would have consumed about 40 kilograms of forage a

Model of a male Irish elk in the
Natural History Museum, London.

day. Furthermore, by the time of the mating season, when the stags
needed their antlers for display and fighting, the antlers had finished
growing, the velvet (skin) had fallen off them, and they were, effectively, dead: so that all their mechanical properties had to be in place
by then. Over the growing months, the stags would have deposited
around 100 grams of calcium into their antlers every day.

This procedure would have been a huge drain on resources. In the
spring, the stags would have been in a poor state, having survived the
thin times of winter – during which males and females separated –
and would have needed to build their whole bodies up again with the
new season's plant growth. The energy expended in calcium-collecting
and storing for the antlers would have represented an enormous
investment. The mystery of why they did it begs another question.
Could it be answered by looking at the behaviour of modern deer?

Adrian Lister, a lecturer in biology at University College, London (and a leading authority on another giant herbivore of the last ice age, the mammoth, as well as *Megaloceros*), has conducted detailed comparisons of the ancient bones and teeth of the Irish elk with those of deer species of today. His findings indicate that the Irish elk is related to the modern fallow deer, still to be found in numbers in, for example, Phoenix Park in Dublin. The fallow deer is, of course, far smaller than its archaic cousin, but the stags have similar broad-bladed, palmed antlers. Lister and his colleagues are currently trying to confirm this theory by comparing the DNA of fallow deer and Irish elk.

Fallow deer employ a system of mating called 'lekking' to attract females. In the courtship ritual the antlers of the males play an important role, and Lister believes that the Irish elk may have behaved in a similar fashion.

In the process of lekking, each mature male selects and lays claim to a small territory, perhaps 100 metres across. That becomes his patch, which he defends against other males. A group of such territories make up a lek. Females visit the lek because once there they have several males to choose from – and it is unusual in nature for the female to have the option of choice. In the influencing of that choice, the antlers play a role.

The antlers demonstrate to the females, by their size, the relative ability on the part of their owners to feed themselves, grow big, and at the same time produce a healthy surplus, which goes into antler-production. The antlers are therefore an honest advertisement of their owner's quality as a mate. During a lek, the broad palms of the antlers, when displayed, flash like mirrors across open landscape, which also serves to attract females to the lek. An additional method of attraction and advertisement was roaring (or 'troating' as it is properly known).

The mating display of fallow deer is not unlike a ritual dance, and arguably Irish elk performed a similar ceremony. For the stag, it is all about ostentation as he exhibits his antler-palms. A large alpha stag in his prime will obviously be more attractive than his fellows younger than five years or older than ten or eleven. One interesting difference in the displays of fallow deer and the Irish elk was pointed out by Stephen Jay Gould: fallow deer have to twist their heads to the side to display their palms, which are directed laterally. Gould suspects that if the Irish elk had to perform the same manoeuvre, it would have encountered severe mechanical problems of torque as it swung its 35-kilogram antlers around on its 2-kilogram skull. Biomechanics expert Andrew Kitchener argues, however, that the force generated by the antlers would have been absorbed by the neck, not the skull. Any twisting, side-to-side movements would be controlled by the muscles, tendons and ligaments of the neck, necessary to support and control the mighty antlers. The skeletons reveal massively developed vertebrae, the dorsal section in particular having huge spines, to which the ligaments would have been attached.

However, the antlers of the Irish elk are so arranged that the full palm is most effectively displayed when the animal is looking straight ahead. All it has to do to show them off is raise its head. This has been noted not only by Gould, but also by scientists Valerius Geist and Russell Coope, each making the observation independently of one another. Coope describes the display as beginning with a slow frontal approach, nose up, lips raised to expose non-existing canines, while hissing and grinding molars. The broadside display would have been reduced in *Megaloceros* to insignificance. The huge, shiny palms (which would have remained very light in colour during most of the rut – as in moose) would have been very conspicuous in open terrain, particularly at dusk and dawn, and would have been well noticed by females. It's possible that very large stags would have been a welcome haven for females from the harassment of young stags – conspicuous antlers would have been adaptive indeed.

Above all, the successful Irish elk stag's antlers demonstrated how healthy his genes were, and so his wooing was successful, and he was able to copulate with the female.

That, however, is not necessarily the end of the story, and nor, according to some naturalists, did the antlers of the Irish elk simply have a display function, unwieldy and even, curiously, delicate as they may seem for any other. The antlers of deer do not consist of palms alone, of course, but also of prong-like projections called 'tines'. Tines have no function in display, but they are useful in attack and defence.

There is a difference in the way various deer hold their heads when fighting, which dictates the angle at which their antlers will

interlock properly. New World deer (including reindeer), for example, hold their heads with the nose pointing directly downwards; hog deer, in common with other Old World deer, tuck their chins in, so that the line of the head forms an acute angle with the neck and their heads are, effectively, even further down. Andrew Kitchener has made a very close technical study of the antlers of the Irish elk, and has been able to demonstrate by 'playing' with two sets of them how the tines were designed to lock together if two stags, heads down, came at one another. Kitchener has based his reconstruction on the fighting posture of Old World deer, and the result is highly convincing.

As the stags put their heads down, rather in the manner of hog deer, the antlers rotated through an arc and the tines were brought into their 'fighting' position, where they could lock together perfectly with those of an opponent, just above the second tine. On impact, the 3-centimetre-thick skull could absorb some of the shock. In other tactics, the huge spread of the antlers was designed to parry the blows from the opponent's attacking tines. The tines above the animal's eyes seem to come down specifically for the protection of that vulnerable area. Darwin described a wapiti, a close relative of the Old World red deer, which, when taking up a defensive posture, 'kept his face almost flat to the ground, with his nose nearly between his forefeet'. Kitchener points out that the antlers of Old World deer, to which group the Irish elk belonged, do not interlock if the head position of New World deer is adopted.

Antlers consist of a thin ring of compact bone around a core of spongy bone – like a foam-filled tube. The closer to the head, the more compact bone there is. The form is designed to take a good deal of stress. The wider the diameter of the 'tube', the thinner the ring of compact bone needs to be to deal with the stress. This minimizes weight and calcium input, while maximizing strength – a perfect piece of engineering. Moreover, the cross-section of the base is slightly elliptical, with the long axis of the ellipse what would have been the main strain of the fighting. The elliptical shape in fact makes the antlers stronger by a factor of 15 per cent. Interestingly, the palm itself had very little compact bone. In effect it is a space-filler between the tines, which themselves are composed of much thicker compact bone. And because locking of tines takes place below the palm, it has to take relatively little stress. This again shows economy: much more material would have to have been used to make a large branching structure than a broad, palmed structure of similar size, and even for an Irish elk it might not have been possible to generate enough calcium once a year to achieve this as well as the fast rate of growth.

Once locked together, the stags would 'wrestle' with their antlers, each trying to get the other off-balance. This method of fighting would exert huge pressure on the base of the antlers, but the long terminal tines grew in a curve aimed directly at the adversary's neck and flanks: if one stag got the upper hand, it could bring these weapons into play to cause serious goring. Andrew Kitchener has calculated that even under the toughest fighting conditions, the antlers would have been unlikely to break – though such an occurrence would not have been disastrous in the long term, since a new set of antlers would be grown for the following year's mating season.

The mechanical properties of antler bone are consistent across all species of deer. Kitchener has also discovered, by comparing the bone of the Irish elk antler with that of a living deer – the hog deer – that, among other things, the internal structure of the antler (the grain) aligns perfectly along the most likely lines of stress if two animals were to lock horns. It is possible to imagine the scene where, during the lek, a hitherto successful stag is challenged by another male, something which, from our knowledge of fallow deer, we can surmise happened regularly. Both animals would raise their heads high at first, to show off their full size and intimidate one another: the body mass behind the charge would count as well as the antlers. They would also bellow – the sound they produced also being part of the intimidation-and-proclamation-of-strength process. Then, if neither backed down, battle would commence, each seeking the advantage of any rise in the ground at the scene of their battle.

Many naturalists believe that battles would not have taken the form of mortal combat. Observations of fights between modern caribou stags reveal that many of them appear to be almost ritualistic. J.P. Kelsall notes that: 'When one bull establishes dominance, by consistently pushing his opponent backwards, the loser disengages, often in an almost casual manner, and walks away.' In the context of giant deer, Valerius Geist observes that populations of large-bodied ungulates have a lower turnover rate than populations of small-bodied ungulates, so in consequence there will be greater selection for non-damaging intraspecific fighting in large-bodied species than in small-bodied species. Broad palms, if they are a factor in dominance, might reduce the need for fierce fighting. But it is also interesting in this connection to note that the number of tines relates more closely to antler width than antler length.

Given all these considerations, it can safely be inferred that antlers were the key to the survival of the species. With them, the Irish elk stag attracted a mate and defended himself against challengers. Antlers could perhaps have played a part in other physical systems, such as thermoregulation. The large investment made in them was essential if he wanted to pass on his genes.

In the case of most deer, this need not be a problem. In the case of an animal as large as the Irish elk, its own massive bulk, coupled with gargantuan antlers, might have played a part in its demise if the climatic and environmental conditions, for which it had evolved, changed. In hard times, did the antlers it sported turn out to be more of a liability than an asset?

Certainly not during mating, as long as the environmental status quo was maintained. To continue the parallel with fallow deer, it follows that the larger, better-armed stags would be able to see off most competitors and thus couple with the most does, spreading their own gene pool as far as possible. In fallow deer, the females tend to live in large groups, and thus it is possible for one male to monopolize large numbers of them, if he is strong enough. (Fallow bucks are more than double the weight of females.) Examples are known of one stag fathering thirty fawns in a year, and anything up to one hundred offspring from one father have been known. This is not the rule for all deer species by any means (roe deer live in a dispersed pattern and males cannot as easily round up, attract or mate with several females) but, as far as we know, male Irish elk were considerably larger than females.

Unfortunately, very few skeletons or skulls of the antler-less, and therefore less collectible, female Irish elk have come to light. Furthermore, several reconstructed skeletons of males and females in museums are not made up of single originals, but are composites, and these are few. Specimen antlers have often been found to have had clever plaster extensions added to them in Victorian times to make them look more impressive. Antlers are rarely found nowadays, and when they come on to the market, collectors have to pay a high price for them. The record set for a pair at auction is £20,000, at Sotheby's in 1989; a complete skeleton changed hands for £27,000 in 1992.

After mating, the males and females would separate, the females staying on high ground, the males descending to the valleys in small groups of two or three to recuperate. The females would form herds of ten to twenty members, perhaps grouped under the leadership of a matriarch. When the young were born, they would have had to be able to run efficiently within a very short time.

The stags, after a long cycle of mating and fighting, with little time to eat, finished the lek exhausted. All their reserves of fat were used up, and at such a vulnerable time came the onset of autumn and colder weather. About 10,000 years ago, as the Ice Age was coming to an end, this need not have been a problem – certainly the stags had survived the changing seasons for millennia; but now an odd thing started to happen: the winters were getting harsher and longer.

It's likely that at around this time the Irish elk started to die in larger numbers, and the radiocarbon dating of bones shows that soon the animal would disappear from Ireland altogether. The climatologist Peter Coxon has been investigating what exactly was happening to the weather locally as the last great ice age came to an end. By pushing a steel tube 8 metres into the ground, using a piston, Coxon was able to take core samples of soil from the period under investigation. From the samples, he saw that there was a gradual change in soil colours as the world warmed up and became wetter at the end of the ice age. But then, like a throwback, the colour changed again: it was just as if Ireland had been hit by a long, terrible winter. This coincides with the disappearance of the Irish elk. In a normal winter, the kind of winters they were used to, a large number of elk would die for want of nutrition, and it was a time when the weak and the old were weeded out; but the winters which came now extended throughout much of the year, and were harder than anything the elk was adapted to withstand.

The die-out is confirmed by excavations. The large number of antlers and bones that have been dug up date to about 10,600 years ago and earlier; after this time, there is no trace of them. In Ireland, man cannot have had a hand in it, since the earliest human settlements date from about 8,000 to 9,000 years ago. No marks of tools, or breaks that would suggest hunting or butchering, have been found on any of the Irish elk bones yet discovered in Ireland.

The disappearance of the elk from Ireland must have taken place very quickly in terms of geological time. Analysed soil samples from this crucial period have revealed more. Pollen extracted from them show that the elk's environment was changing quickly and radically with the climatic change. The rich grasslands were disappearing, and calcium-yielding plants like willow were growing scarce. Suddenly, the elk was stripped of the key elements of its diet. It should be added that when the Irish elk had reappeared in Ireland about 12,400 years ago, Ireland was already beginning a vegetational change from dwarf willow scrub to a juniper and crowberry scrub. This then gave way to grassland in which both willow and juniper were scarcer than they had been. It is possible that the Irish elk itself, together with the reindeer, which also lived in Ireland in great numbers at that period, may have influenced the change themselves through overgrazing. The dominant reason for the decline of the grasslands from about 11,000 years ago, however, was the drop in temperature. It is interesting that reindeer vanished from Ireland at about the same time as the giant deer.

Why should the local weather suddenly turn so dramatically much colder at the very end of the ice age? The answer lies in a climatic effect now known as the Younger Dryas period. With general global warming, the frozen northern glaciers were melting, dumping a huge volume of cold water into the ocean, and nullifying the effect of the North Atlantic Gulf Stream, which accounts for the British Isles' temperate climate even today. With this heating system switched off, the mean temperature dropped by seven degrees Celsius over the next twenty years – enough to plunge Ireland into a permanent winter. The Irish landscape quickly came to resemble tundra. The ground became frosty and cracked.

Smaller mammals stood a better chance of adapting to such conditions, but animals of huge bulk, which needed a massive daily intake of food, were placed in a highly vulnerable position. Instinctively they would have followed the usual cycle of the seasons, but the demands this made upon them physically now proved too great a debt for them to pay. The late autumn and winter weather conditions that followed the mating season were especially tough. Food was hard to find, and relatively weaker deer had to travel relatively further to find relatively less of it than before.

Under severe nutritional stress, the Irish elk may have developed osteoporosis – hollow bone syndrome – as it devoted precious calcium surpluses to its antlers, even though its nutrition was suddenly deteriorating. In the adverse conditions of the Younger Dryas, creating the antlers would have cost a great deal.

In terms of traditional evolutionary theory, survival in the long term could be guaranteed only by the evolution of smaller antlers, but the antlers' size did not decrease. Such a lack of adaptation seems to fly in the face of nature; but sexual selection is a primal force, and Irish elk does did not change their genetic programming to react to large antlers and choose their mates according to their antlers' dimensions. Millennia of evolved behaviour could not be unlearnt

quickly. In order to mate and pass on their genes, stags were obliged to continue to grow huge antlers. The tragedy for the Irish elk was that to survive it would have to adapt, but the sexual imperative was against that.

Caught in a vicious circle, there was no escape from extinction for the Irish elk. However, the previously accepted date of its demise, around 10,600 years ago, may now have to be revised in the light of recent discoveries made on the Isle of Man by Adrian Lister and a colleague, geologist Silvia Gonzales of the University of Liverpool (whom we last encountered in the previous chapter examining mammoth finds).

Lister and Gonzalez were excavating a kettle-hole – a glacial hole in the landscape that dates back to the ice age and has since been filled in with soft soil and plant debris. Storms had weathered away the sides of the kettle-hole, exposing soil and the fossils that have collected in it for thousands of years – and here Lister and Gonzalez uncovered the remains of Irish elk. This was not the first time that the giant deer had been found on the Isle of Man, where it also lived. An almost complete skeleton was discovered here as early as 1819. Recently, however, modern examination of this skeleton, and radio-carbon dating of it by Lister and Gonzales, revealed that the animals here lived until around 9,200 years ago – 1,400 years later than they were supposed to have become extinct, and indeed did apparently die out in Ireland. Tom Hayden has likened this discovery to the equivalent in historical terms of finding that the Mayan civilization was still thriving today on an offshore island.

This was possible because, when the Younger Dryas period set in, the Isle of Man was still connected to the British mainland. At the time, Ireland was not, because the Irish Sea is very deep and to the west of Man no land bridge was formed even during an era when the seas were generally lower than they are today. However, both the Isle of Man, and, to a greater extent, southern Scotland, were more protected from the cold effect coming from the Atlantic Ocean than Ireland was. During the winter months, the elk living on Man could cross over to the mainland where they may have found the nutrients they needed to survive. An antler fragment found near the River Cree in south-western Scotland has been dated to 9,340 years ago, and a nearly complete skeleton from Loughan Ruy, Ballaugh, on the Isle of Man, to 9,225 years ago.

The scientists further discovered that the fossil bones discovered on Man were smaller than those in Ireland; by comparing skull sizes they found that the Man deer was about 10 per cent smaller than its Irish counterpart. Over the 1,400 years from the onset of the Younger Dryas, they had evolved smaller bodies, which required less food to sustain. However, although one might have expected the antlers to have decreased in size as well, those unearthed so far have been just as big as ever, so that, relative to body size, they were even larger. This indicated that sexual selection was still putting pressure on the stags to produce large antlers: the animal was able to adapt in order to survive, but not as far as its antlers were concerned.

Although absence of fossils is always negative evidence and could be contradicted, the existing fossil record suggests that after 9,200 years ago the Irish elk disappeared even from its ranges on the Isle of Man and in Scotland. Warmer, wetter weather asserted itself

again as the effect of the Younger Dryas waned about 10,000 years ago, and this meant a further change in the environment. The forests that had pre-dated the ice age returned. The Irish elk had evolved for a life in open or semi-open terrain. It was poorly adapted to life in a dense forest: the huge spread of its antlers would have hampered it among trees (woodland-dwelling deer have smaller antlers that sweep backwards in line with the body), and the 'new' weather was producing plant life that was unsuitable for its dietary requirements. But there was something else too. If the Irish elk had been able to survive the Younger Dryas, it would have lived long enough – arguably – to have encountered the first Mesolithic human pioneers migrating north across Europe with the warmer weather.

It is known that the skulls, with antlers attached, of red deer were adapted for religious or ritualistic purposes by Middle Stone Age man. Intricate headdresses and masks were fashioned from the antlers and skulls, and those of red deer were highly prized. If humans encountered the Irish elk, how much more valuable would its antlers have appeared to them, both as prized possessions, conferring status on the owners, and as symbols of authority? We know that man encountered the Irish elk from 25,000-year-old cave paintings at Cougnac, in south-western France. In these, the animal is clearly drawn in a natural pose, looking somewhat eland-like, the head held much lower than is suggested by the mounted skeletons in museums, where it is usually held uncomfortably high, to show off the animal's size. The paintings at Cougnac were discovered in 1952, and show two males and a female. They indicate distinct black lines from the shoulder hump stretching diagonally down the flank towards the back legs, but it is not known whether they indicate markings or shadows. Unfortunately the colour of the animals' hide cannot be deduced indisputably from the paintings, but it would probably have been light, to reflect the sun – pale brown or dark yellow. Clearly the artists were proud of their work: several stalactites have been broken away to improve the view of them, and they are placed in an optimum position for viewing. Although giant deer are depicted in cave paintings in four other French sites, they are otherwise rare, which fact has led some scientists to conclude that the Irish elk was itself a rare animal, at least in those areas.

It is known that humans had arrived in Britain between 40,000 and 25,000 years ago, though no evidence at all of any very early encounters between man and elk so far north have yet come to light.

What little evidence there is of the culture of Mesolithic Man in England comes from a dig conducted in the 1950s at Star Carr in Yorkshire. A re-examination of fossil bones found there may reveal whether or not the Irish elk was ever pursued by Stone Age hunters. So far, no Irish elk bones have been discovered but, as more Mesolithic sites are identified, the evidence needed to support such a theory may also be revealed.

In Scotland, on the banks of the River Clyde, another site has been uncovered by the archaeologist Tam Ward. It dates back to 10,000 years ago – far earlier than anyone had supposed humans to have migrated this far north. Here, it is certainly possible that man and the Irish elk co-existed, perhaps for hundreds of years.

Mesolithic Man differed from his Palaeolithic forebears in that for the first time he made an appreciable impact on his environment. He burned down trees and even swathes of forest to flush out prey, and hunted from a distance for the first time, having invented the bow and arrow. If he encountered the Irish elk, Ward thinks it likely that he could have hunted it with success. The coincidence of man's arrival with the final demise of the Irish elk suggests a theory which seems to make sense. Furthermore, Mesolithic Man deliberately camped on hill sites and other places well suited for the observation and hunting of deer. The site Ward discovered, for example, a hill bivouac, was placed so that the hunters who camped here could look into three valleys at once.

Perhaps the combination of unsuitable habitat and a new predator dealt the coup de grâce to the already doomed Irish elk, but for the moment at least it remains a matter for conjecture.

LOCATIONS

We needed to film landscapes in order to produce what computer animators call backplates. In effect, they're the stages on to which the computer animal will be placed and animated as if it were moving through a real environment. No matter how realistic the computer-generated animals were, the impact would be lessened if the landscapes they were put into weren't convincing.

For three of our animals – the Irish elk, Columbian mammoth and sabre-toothed tiger – we needed to re-create the world as it was at the end of the last ice age. But it wasn't merely a question of going somewhere cold. Each of these animals lived in different parts of the world, with different climates, vegetation and scenery. The fossil evidence placed these creatures in particular habitats. The sabre-tooth is best known from finds in the La Brea tar pits, in what is now the heart of urban Los Angeles. New evidence about mammoths comes from the site of Tocuila, which lies close to Mexico City – another vast

A working ranch, complete with bison.

Filming the sabre-tooth
animatronic on location.

urban sprawl, in the basin of Mexico, in a giant valley surrounded by
volcanic peaks.

Ten thousand years ago, before these huge cities existed, the vegetation and climates were different, although the basic scenery
remained the same. We knew from our research that we'd be able to
shoot both the mammoth and sabre-tooth backplates in America.
The challenge was: where, exactly?

Southern California used to be wetter than the semi-arid
Mediterranean-style climate it's known for today. There were lush
grasslands and stands of trees along rivers in which the sabre-tooth
stalked. These conditions can still be found in the area close to San
Francisco, near Monterey. Photographs looked promising, so a local
location finder (of which there are many in California) was
dispatched to investigate.

One added complication in finding locations for the sabre-tooth
film was the need to shoot backplates that contained living animals,

specifically the North American bison. Director Jenny Ash was writing a scene in which the sabre-tooth hunted and killed a bison, and other scenes in which our computer-generated predator was seen in the same shot as the real animals. Unlike the sabre-tooth, the bison survived the ice age and continued to roam the western part of the continent in huge numbers. To save having to build a whole herd of computer bison, which would never look as good as the real thing, we decided that it made sense to film the scenes in which bison interacted with their long-extinct enemy using real bison. We chose the V6 ranch, near Monterey, where a herd of bison roamed land that looked much as it would have 10,000 years ago.

Back in Europe, the Irish elk roamed the hills and mountains of Ireland, and we initially thought location hunting would be comparatively simple. Ironically, finding a place to shoot the backplates for the Irish elk proved very difficult. Ireland's landscape, like the British Isles as a whole, is heavily inhabited and farmed. Since prehistoric

The island of Eldey, the last home of the great auk, thirty miles off the coast of Iceland.

times, humans have been settling, building and raising their animals, which has changed much of the countryside beyond all recognition. Compared to North America, where vast wilderness areas still remain, few places in Ireland have evaded the hand of man. Luckily, there are a few places which remain largely untouched. One of these is the spectacular Glen Veagh national park in County Donegal. Glen Veagh is rugged mountain country, incised by deep glacial valleys and a large lake, Loch Veagh. It was a perfect place, a landscape created when ice sheets covered the northern part of the country.

However, foot and mouth disease broke out in the UK in late February 2001, and farmers in Ireland were concerned that the disease should not spread there. In Britain, the number of cases diagnosed rose week by week, and the public were advised not to enter the countryside unnecessarily. Even so, the disease was reported in Armagh in Northern Ireland by early March. With preparations for the shoot in Glen Veagh in full swing and assistant producer Dan Hillman in the middle of a recce to fix the locations we would use, the park took the reluctant decision to close, with no shooting for the foreseeable future.

Actors in period costume filming a reconstruction of fishermen rowing to Eldey.

Foot and mouth disease also impacted on another film in the series. We'd been exploring the nearby Donegal coast for backplates for the great auk film. It is rocky and inhospitable, and provides a good match for the remote rocky coastlines of Newfoundland and Iceland where much of the action in our story took place. On top of that, Donegal would be much less physically gruelling to shoot in. But with Ireland out of bounds, we quickly decided that Iceland was the best place to go. At least we'd have no problem getting the right kind of landscape there. It was now a question of finding the best-looking location that was also safe and accessible. It had to be close enough to civilization to prevent cast and crew having to carry themselves and a lot of heavy equipment over hill and dale, yet rugged and wild enough to be convincing.

The Irish elk proved a more intractable problem, however. Ireland is a very distinctive looking country, with its green valleys and rounded mountains. We'd been looking at Oregon, one of America's most beautifully wild and magnificent states, for the Columbian mammoth, and through the Oregon Film Office we ended up at

The crew on location off the Icelandic coast.

Klamath Falls. We filmed in a wide valley, surrounded by ranges of jagged, snow-capped peaks. Pine trees added to the sense of sub-glacial, semi-arctic environment – how Mexico probably looked in the ice age.

That left just two creatures to find backplate locations for: the dodo and the Tasmanian tiger. Since both species died out compara-tively recently, we didn't have to find locations to resemble long-changed environments. In addition, since both animals lived on specific islands, there was no need to search any more widely than the places they'd actually inhabited.

Nothing's ever that straightforward, of course, especially in the case of the dodo, as it turned out. This remarkable bird evolved on the island of Mauritius in the Indian Ocean, but it had done so in a tropical landscape then untouched by man. Since the Dutch discov-ered the island, and the bird, in 1598, all that has changed. Modern Mauritius has been taken over by a plant the European colonists brought with them to farm – sugar cane. Sugar cane has created a virtual monoculture where lush jungle and forests of ebony once stood. Finding the sort of habitat that the dodo thrived in prior to the arrival of man wasn't looking as easy as we had first imagined.

While the interior mountains are still forested, expert Julian Hume cautioned that not just any forest would do. The habitat had to be what he called 'Mauritius Dry Forest', the homeland of the dodo. Historically, it had existed on the eastern coastal plain of the island, growing low enough to avoid the mountain rainfall that created thick tropical jungle. It was a unique habitat with relatively open forest floor under the tall ebony trees. Ideal for ground-dwelling dodos to nest in.

With Julian as guide, director Russell Barnes and Dan Hillman went in search of this particular habitat, and found a small patch in a forest reserve on the south side of the island. They set about trans-forming it into the arena in which the computer-generated dodo would live, eat, mate, and ultimately face extinction.

With the choice of locations now resolved, the major task of actually filming the backplates, animals and actors was ready to begin. The *Extinct* production team began to try to realize the vision of bringing six long-dead species and their worlds back to life.

DODO

Raphus cucullatus

For the fleet, it might have been a disaster. It was a large expedition to the East Indies from the Netherlands, involving eight ships under the command of Admiral Jacob Corneliszoon van Neck – an expensive investment for the syndicate of Dutch merchants that had backed it. But scarcely had they rounded the Cape of Good Hope on their outward voyage than they ran into heavy weather that split them into two groups. Five of the ships sailed east of Madagascar and made landfall only when their water casks were running dangerously low. They were pretty sure of where they were – the lush, mountainous island that rose before them out of the sea had already been identified on old Portuguese charts as the Islo do Cerne (Island of Swans). The problem was finding a safe harbour. Finding one was a matter of urgency: not only were they short of fresh water, but scurvy had taken hold among several of the crews, so that the need for fresh food, and an opportunity for the sick to rest and recuperate, was urgent. The green island seemed like a gift from God. It was 17 September 1598.

On the following day, sloops were sent out from the ships in search of a safe anchorage. At first none was found, but then two sloops, one from the *Amsterdam* and the other from her sister ship, the *Gelderland*, hit upon a natural haven that they named Warwijk Harbour, after their vice-admiral, Wybrant van Warwijk, in command of the *Amsterdam*. When the crews went ashore – the men of the *Gelderland*'s sloop landed first – they encountered what seemed to be a paradise on Earth.

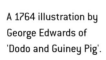

A 1764 illustration by George Edwards of 'Dodo and Guiney Pig'.

Palm trees grew close to the shore, and beyond them were thick forests of ebony. The men did not explore far on that first day, but they took note of a variety of birdlife. Pigeons and parrots were plentiful, but there was another kind of bird there too – a big bird, apparently unable to fly. None of the wildlife they saw had any fear of man, and there was no sign either of other creatures or of human habitation. The crews of the two sloops found that they could pluck the pigeons from the trees. They managed to seize nine of the large birds too, grabbing them by their stumpy wings

A drawing by Alfred Waterhouse for a panel in the Natural History Museum, London, *c*1875.

and hauling them into the boats, taking care to avoid the long beaks that ended in a businesslike hook – though the creatures, never having encountered man before, seemed passive enough, according to the earliest reports. That evening the crews returned with their booty – and casks of fresh water, which they also found in abundance – to their mother ships.

On 19 September, guided by the sloops, the five large ships sailed safely into the harbour. The men disembarked and held a service of thanksgiving on the shore. It was the day on which the annual fair – the *kermis* – was held at home in Amsterdam, so they decided to call the curious large fowl that emerged from the trees to observe them the kermis-goose. An early description of these birds speaks of a creature 'as big as our swan, with large heads, and on the head a veil as though they had a small hood on their head; they have no wings but in their place there are three or four black quills, and where there ought to be a tail, there are four or five small curled plumes of a greyish colour'.

The name kermis-goose might have stuck – after all, the bird, though it squawked loudly and struggled when caught, was easy enough to overpower, and its size promised good eating. It was only after they killed a few of them, bludgeoning them to death with staves, and tried to cook them, that the sailors discovered the likelihood of a feast looked thinner. Although the breast-meat tasted good, there was very little of it, for the birds, despite their size, did not have the plumpness of a goose. Early reports tell that the 'stomach' was good to eat too, but as for the rest, no matter how long they stewed the meat, the sailors found that it remained tough and sinewy, hard to chew (especially with teeth loosened by scurvy), and oily and unpleasant to the taste. A new name was quickly found for the bird – *walghvogel*, 'nauseating bird'. The Dutchmen turned their attention to the pigeons, which were far tastier and much easier to cook, and to the fish that teemed in the harbour. To their disappointment, they found no other animals to eat, though there had been rumours that decades earlier the Portuguese had released a few pigs on the island.

It would not be long before the Dutch coined yet another name for their new discovery. The bird, turkey-like, stored fat in its rump. From that the name *doedaars* sprang – 'fat-arse'. Given the fact that the ships' crews came from several European nations, and that international interest in the curious and exotic creatures grew quickly after the first reports got back to Europe, it is not surprising that this name was quickly corrupted to 'dodo': the name that stuck.

Those five Dutch ships (which would ultimately be reunited with the rest of the expedition in the East Indies) stayed in Mauritius for just twelve days – but that was long enough to seal the dodos' fate. Well before the end of the next century, it was extinct. For a long time it was assumed that it was the victim of its own docility – even stupidity – and that it was quickly hunted out for its flesh. But the early reports show that as a source of food it was most unpopular, and recent research has called into question the strongly held view established in the nineteenth century that the bird was fat, slow and dim – an obvious victim of natural selection. So what really happened to the dodo, and why did it vanish from the Earth within seventy-five years of its discovery by man?

The crews of the Dutch expedition were probably the first human beings ever to see a dodo. Mauritius itself had been discovered as early as 1500, by Portuguese ships in quest of a route for the East Indies. Along with Réunion and Rodrigues, discovered later in the century, it was part of the Mascarene Islands. Given the Portuguese name for Mauritius – Island of Swans – it is possible that a dodo had been sighted at some point, and mistaken by those early explorers for a swan.

Portugal, one of the great cradles of maritime exploration, was a small country with a small population. Its principal colony was Brazil, brokered through the Treaty of Tordesillas with Spain in 1494. Brazil was still virtually virgin territory beyond the coastline, and Portugal had neither the resources nor the manpower to exploit the far-flung territories its sailors discovered in their attempts to develop Portuguese ambitions in the East Indies (the Spanish having laid claim through the same Treaty of Tordesillas to all lands in the western hemisphere). Furthermore, although Portugal had enjoyed independence and control of the southern and eastern seas for most of the sixteenth century, in 1580 – when the direct male line of its kings came to an end – Philip II of Spain annexed the country despite the efforts of the English to support a Portuguese pretender.

But the Spanish had their hands full. Their resources were at full stretch managing their empire in South America and the West Indies, and their vassals in Europe did not lie easily under their yoke. Spanish suppression in Portugal led to a revolt in 1640, which established a Portuguese king once more, though the country's overseas interests were in decline, and it would be another thirty years before the country achieved full and peaceful independence. In the meantime, the other European vassal, the Dutch United Provinces (the modern Netherlands), as great a maritime nation as Portugal, Spain and England, had been engaged in a similarly long struggle for independence. Its war with Spain lasted from 1568 to 1648. By 1585, supported by England, the Netherlands had achieved autonomy, and under Maurits of Nassau began to develop their own colonial empire.

It is not surprising that the Dutch discoverers of the dodos' island named it Mauritius in honour of their prince. As early as 1598 the individual Dutch merchant syndicates combined to establish the East India Company, which mounted two expeditions in 1600. In the same year, the

A dodo skeleton in the Oxford University Museum of Natural History.

company merged with the New Brabant Company to form the First United East India Company of Amsterdam – the forerunner of the Dutch East India Company.

In 1609, the conclusion of a twelve-year truce with Spain enabled the Netherlands to expand their own interests in the East Indies further. 1621 saw a renewal of the struggle with Spain, but at last full Dutch independence was recognized by the Peace of The Hague in 1648. By then, however, the Netherlands was well established in the East Indies. Mauritius became a port for rest and recuperation after ships rounded the Cape, before heading across the Indian Ocean to the spice plantations. The abundance of ebony had not escaped the attention of the Dutch either, and logging would begin within a few decades, though the settlement – or 'factory' – remained small, and there was no serious colonization of the island. The sailors, however, found it expedient to release pigs, which thrived and bred plentifully, to ensure a good supply of palatable meat. On the ships with the pigs came other interlopers – cats, rats and monkeys.

There had been no such creatures on the island before, nor had Mauritius ever been inhabited by man. The Portuguese had charted it, but both Portugal and Spain kept the maps they drew a closely guarded secret; if any charts exist from the Arab world, they are yet to come to light. Spanish 'rutters', or sea-maps, were occasionally seized by pirate or privateer ships and brought back to their home countries, and these were highly prized. With the development of the printing trade, charts and maps became ever more readily available, and it became increasingly difficult to keep routes and islands on those routes secret. More and more ships were sailing on voyages of exploration throughout the seventeenth century, and superstition about what might be encountered on such voyages quickly gave way to a spirit of scientific (and mercantile) enquiry. The sixteenth century had been a kind of watershed between the late Middle Ages and the dawn of the Age of Reason.

In addition to the maps, ships' logs carefully recorded, for the sake of those who came after, exactly how to find moorings and what to expect on land – as the report on Mauritius, from the 1598 voyage, testifies. No edition in Dutch remains, but – indicating how fast news travelled and interest grew – in 1599 an English version appeared, from which this quotation comes:

> This Island being situate to the East of Madagascar, and contain-ing as much in compasse as all Holland, is a very high, goodly and pleasant land, full of green and fruitful vallies, and replenished with Palmito-trees, from wich droppeth holesome wine. Likewise here are very many trees of right Ebenwood as black as jet and smooth and hard as the very Ivory; and the quantity of this wood is so exceeding that many ships may be laden therewith.

> For to sail into this haven you must bring the two highest moun-tains one over the other, leaving five small islands on your right hand; and so you may enter in upon 30 fadomes of water. Lying in the bay they had 10, 12 and 14 fadoms. On their left hand was a little island which they named Hemskerk Island…

Mauritius is a rugged, hilly island of some 1,865 square kilometres, surrounded by coral reefs. It has a heavy rainfall and tends to be hot and humid. The Mauritian summer corresponds roughly to the European winter, and vice versa. Nowadays the ebony is gone, and the principal crops are sugar, molasses, copra, rum, aloe fibre, coconut oil and vanilla. Apart from Réunion and Rodrigues, each a sizeable distance from it, it is isolated in the ocean 850 kilometres from Madagascar and 3,900 kilometres north-east of the Cape. On Mauritius the dodo lived in equally splendid isolation and with great success. But how did it get there, what was its history, and how did it react to man's invasion?

Mauritius is approximately 10.5 million years old. Rodrigues is about 8 million years old, and Réunion about 3 million. Everything on Mauritius is also on Réunion, apart from the dodo – which means that it must have been flightless before Réunion was formed.

Natural history as we understand it is a relatively new science, and the reason we know so little about the dodo is that in the short time left to it following its first encounter with man no serious attempt to document it was made. Exotic animals discovered and brought back – dead or alive – to Europe in the course of burgeoning

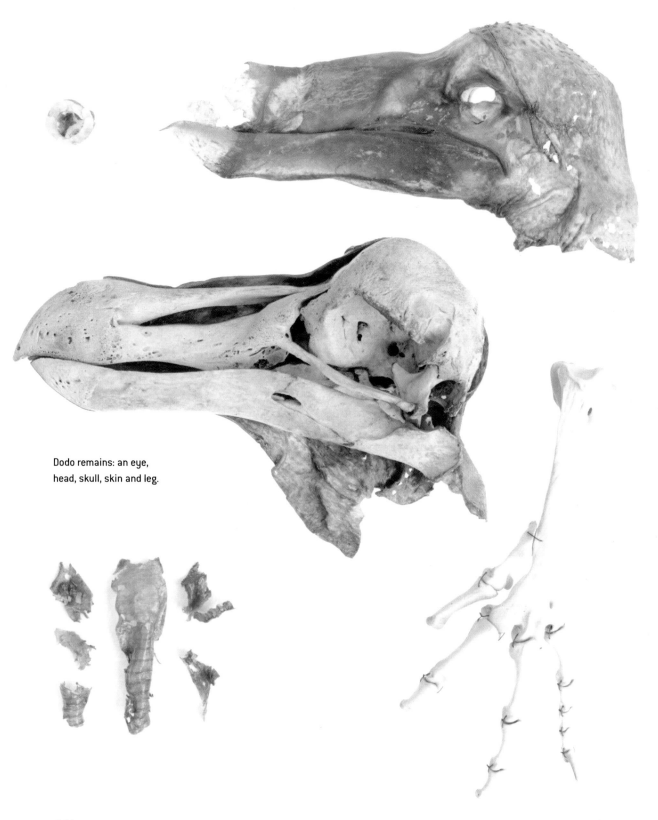

Dodo remains: an eye,
head, skull, skin and leg.

DODO: VITAL STATISTICS

Raphus cucullatus (Linnaeus, 1758)
– also (previously) *Didus ineptus*

Common names: dodo, doddaers (various sp.), dronte, kermis-goose, 'nauseating bird'.

Size: large, comparable to swan, goose, turkey.

Weight: *c.* 10–22 kg (females slightly smaller and lighter than males).

Posture: probably upright, breast held high.

Presumed coloration: body mousy to dark grey, neck darker, bill light with darker coloured tip, very large, opens extremely wide; face naked of feathers, light grey skin, feathered hood; legs and feet yellow, talons black.

Plumage: sparse, downy, with exception of more developed tail and wing feathers.

Nesting: ground nest of leaves and grasses in woods on Mauritius and its islets; one white egg once every two years the size of a pelican's egg. Incubation time uncertain.

Behaviour: unafraid, acquired aggression/fleeing when confronted by man.

Call: gosling-like (contemporary reports) or perhaps pigeon-like. Loud squawk (reported) when alarmed.

Food: fruit, leaves, invertebrates, coral.

seventeenth-century maritime exploration were regarded more or less as curiosities. They were wonderful, half-mythic creatures that belonged more to the realms of sea-serpents and chimeras than to the animals of the then known world (and those of Africa and Asia, which became known in ancient and medieval times through the development of overland trade routes and the expansion of the empires of antiquity). In any case, maritime exploration was chiefly concerned with political and mercantile advantage, not with the fauna and flora of the new territories, except where these lent themselves to commercial exploitation.

The 1598 expedition was quickly followed by others, and several journals by those involved survive. Although the very first report describes the dodo as twice as big as a swan, the next more accurately records that it was about the same size as a swan. Others are more or less consistent: as big as two penguins, as big as a lamb, and so on. Many note that the birds were hard to cook, and unsavoury except for the breast and the stomach. The early journals note that dodos were docile, and easy to catch and kill.

Some sailor artists drew the dodo in its native habitat. One of them, Joris Laerle of the *Gelderland*, has left impressions of a thin, muscular bird; but these contrast interestingly with other drawings of the dodo in captivity – the fat, ungainly creature we know from Sir John Tenniel's illustrations for Lewis Carroll's *Alice in Wonderland*, and from the famous nineteenth-century reconstruction of the bird in London's Natural History Museum. But what is the true picture? It is certainly obfuscated by the fact that some representations of the dodo have given it webbed feet, which it did not have, and even two right feet. Confused or exaggerated reports when the bird was alive added to the confusion. Only now, over 300 years since the dodo's demise, is science able to piece together what it was really like.

The dodo was certainly not dim. With time, man learned a little bit more about the bird, but the bird quickly learned that man was a vicious interloper and adapted its behaviour accordingly. When man arrived, the dodo population must have been plentiful in relation to its environment, because so many were killed so easily – one account records thirty birds taken within an hour. Why this should be the case, given that the birds were bad eating, can only be attributed to brutality. Other wildlife suffered the same fate. Already by 1602 Captain Willem van West-Zanen records in his journal that his men killed '24 or 25 Dod-aarsen, so big and heavy that scarcely two were consumed at meal-time, and all that were remaining flung into salt'.

In 1613 the journal of Johan Verken, a German sailor from Leipzig (professional sailors had a higher-than-average level of literacy), was published in Frankfurt-am-Main. He writes of the birds on Mauritius, including a new description of the dodo. Verken records more concisely than previous chroniclers the dodo's featherless 'face', and for the first time gives the colours of its plumage – yellow feathers at the tips of its stumpy wings, and 'in place of a tail four or five curly greyish feathers'. Even more significant is his description of its behaviour – no longer does the dodo accept its fate complacently. 'They are known as Tottersten or Walck birds. They are found… in large numbers, though the Dutch have been catching and eating them daily, and not only these birds, but many other kinds… which they beat with sticks and catch, taking care all the while that the Tottersten or Walck birds do not bite them on the arm or leg with their great, thick, curved beaks.'

The dodo was obviously, within thirteen years of being introduced to man, not going to take things lying down. It had never known a predator before. It had been king of its own castle. Now that an enemy had come from the sea, it quickly learned to defend itself, using the powerful bill with which males (possibly) saw off rivals during the mating season.

The first man to mention the dodo by that name and spelling in print was Thomas Herbert, an English diplomat to the Persian Court who found himself in Mauritius on his journey home in 1629. He published a book in 1634 recounting his travels, and over several editions elaborated his descriptions, including his observations of the dodo. He gives the bird a melancholy countenance, a bright eye like a diamond, and a detailed portrait of the head: 'The half of her head is naked, seeming covered with a fine vail, her bill is crooked downwards, in the midst is the thrill [nostril], from which part to the end is a light green, mixt with a pale yellow tincture.' The green spot may have been there to guide the young, who would tease the parent's beak open in order to induce it to disgorge food.

Herbert writes that the dodo weighed fifty pounds, that it had a big appetite, and that it had a 'fiery' stomach, capable of digesting anything, even stones. This last was a confusion between the bird's habit of swallowing stones for its gizzard, and its notional ability to consume the hardest materials. Herbert was wrong in several particulars, but his general description of the bird is consistent with what we already know, and adds useful details. His description, along with those of others, combines memories of such birds as the ostrich and the cassowary – both of which shared the dodo's reputation for a supernatural digestive ability. We have to piece together fragile and subjective evidence from eyewitnesses. Only recently have scientists been able to make objective palaeontological reconstructions.

By 1611 the hitherto peaceful, fruit-eating dodos may have learned to bite back, but they had not yet understood that they might also have to flee. 'They displayed themselves to us with a stiff and stern face and wide open mouth,' writes another sailor in 1631, 'very jaunty and audacious of gait, and would scarcely move a foot before us.' If you could avoid being pecked, you could still catch yourself a dodo. Not all of them were killed. A particularly observant diarist and traveller, Peter Mundy, kept a travel journal between 1628 and 1634 in which he notes having seen two dodos kept at Surat, where the Dutch East India Company had a station, and where Jahangir, the Great Mogul of Persia, kept a menagerie. Mundy recalled seeing dodos in Mauritius as well, and is the first man to point out that they seem to occur nowhere else in the wild. He also mentions that they can 'neither flye nor swymm, being Cloven-footed', and speculates about how they could ever have reached so remote an island. In captivity, dodos were taken as curiosities to Amsterdam, Prague, Batavia, Nagasaki and London. It is a testimony to the toughness of the dodo that it was able to survive such journeys.

The London dodo was seen by one Sir Hamon Lestrange, who wrote of it:

About 1638, as I walked London streets, I saw the picture of a strange fowle hung out upon a cloth... and myselfe with one or two more then in company went in to see it. It was kept in a chamber, and was a great fowle somewhat bigger than the largest

Turkey Cock, and so legged and footed, but stouter and thicker and of a more erect shape, coloured before like the breast of a yong cock feson [pheasant], and on the back of a dun or deare colour.

The keeper called it a Dodo, and in the ende of a chymney in the chamber there lay a heape of large pebble stones, whereof hee gave it many in our sight, some as big as nutmegs, and the keeper told us she eats them (conducive to digestion)…

Lestrange's description goes on to tell of the bird's erect posture, and its proud bearing. There are also beginning to be more clues about its general appearance: young pheasants are a chestnut brown, and 'dun' suggests a colour approaching the fur of a house mouse. Earlier descriptions of the bird's colour concur: 'ashy grey with a white face', says van Neck. The eye was usually described as bright, with white irises that must have enhanced the impression. The legs were yellow and the talons black.

Lestrange's account is detailed and rational. Unfortunately, not all contemporary memoirs of the dodo lead us towards further enlightenment. François Cauche visited Mauritius in the same year that Lestrange saw his dodo in London, but from what he writes one has to sift new relevant detail from a mélange of other information, imported from confused memories of other birds. Cauche states that the dodo has no tongue – but that is a mythical characteristic of the cassowary. On the other hand, we do learn that the dodo lays a white egg, and one only, 'the size of a penny bun' – which comparison is not enormously helpful unless one knows how big a French penny bun was in 1638. And in fact Cauche himself saves speculation by telling us, later in his book, that the egg of the pink pelican was the same size. This bird is still with us, so we can infer that a dodo's egg was about 8 centimetres by 6. From Cauche we also learn that dodos made their nests at ground level on a pile of grass – which seems entirely reasonable, given that before man's arrival dodos had nothing to fear – and that they make their nest 'in the woods'. Cauche also tells us that the dodo had a call 'like a gosling', in other words, a kind of soft cooing; that its tail feathers were the same in number as the bird's own age (something which it is impossible to confirm now);

and, significantly, that its feathers overall were downy. Being flightless, the dodo had no need to develop sophisticated feathers, so they remained at a juvenile stage. But, as will become clear, its size was a measure of its success.

Various other memoirs of the dodo crop up as the seventeenth century progresses. It is interesting that as early as the 1630s Peter Mundy reports ominously that when visiting Mauritius a second time, seeking the dodo, 'we... Mett with none' – but the first and last indication that the bird was in deep trouble comes from the journal of a shipwrecked German sailor. On 11 February 1662 a ferocious storm overtook a fleet of seven ships east of Madagascar. Three vanished without trace. Some of the large crew of the *Arnhem* managed to get away in a sloop and, after suffering terrible privations, Volkert Iversen and eighty other survivors managed to beach at Mauritius.

Their voyage after the storm had been appalling: blown off course, they had had to throw the sick and the injured off the overcrowded boat; others went mad from drinking seawater or their own urine. As the situation became increasingly desperate, the officers aboard proposed the plucking of straws to determine who should be eaten first. At last – mercifully before they needed to resort to cannibalism – the survivors were thrown on to the Mauritian shore.

Soon, eleven of their number, including most of the officers, sailed away again in the repaired sloop in search of help – help that never came. Of those who remained, four would write accounts of their adventures. The German sailor, Volkert Iversen, who spent six months on the still uninhabited Mauritius, gives an account of the last days of the dodo.

By the time Iversen arrived, only sixty-four years after the *Gelderland* sloop made landfall, there were no dodos left on Mauritius. The only place where they could still be found was on the islets offshore, where the stricken creatures had retreated via a sandbar at low tide. With four companions, Iversen went hunting there:

We also found here many wild goats and all kinds of birds which are not at all timid, perhaps since they are not used to seeing people who hunt them. They stood quite still and watched us and allowed us to approach them. Among them were the birds known to the Indians [sic] as Dodderse, which are larger than geese, but unable to fly, having only little stumps of wings, but are fast runners. One party of us would chase them so that they ran towards the other party, who then grabbed them. When we had one tightly gripped around the leg it would cry out, then the others would come to its aid and be caught as well.

So by this time dodos had learned to run away from man, and run fast – evidently the only way sailors could catch them was to round them up. But it was too late for the dodo. No mention is made of any left on the main island, and three companions of Iversen who also kept journals scarcely mention the bird at all. Iversen was the last man to see a living dodo and record the fact, and by the time he saw them their population had dwindled past any point from which it could ever recover.

We are left with conflicting impressions of the bird – we cannot take contemporary reports at their face value because they are unsupported by truly scientific observation, though there is enough consistency in the early descriptions to suggest a coherent picture. The inconsistency arises between the energetic and intelligent dodo that emerges from sailors' memoirs and drawings from Mauritius, a bird that furthermore is not necessarily even fat, and the dodo of the popular imagination – plump, docile and dim. But in illustrations of the bird there are large inconsistencies, and even its posture is not recorded in the same way: some show it upright, like a penguin, others give it the general outline of a turkey. It seems to have walked with a stiff-legged gait, but it could move very fast when the need arose and, for its general shape, it was agile. For a long time people assumed that the dodo must have been some kind of sea-bird: otherwise how could it have got to Mauritius?

Pictures showing the fat bird were drawn of the dodo in captivity. After a long sea voyage and then a lifetime kept in a smallish pen and possibly either overfed or given an unsuitable diet, the bird might well have put on weight and developed lazier habits than it would have had in the wild. It is these captive dodos that fired the popular imagination and gave rise to the myths, until recently accepted as facts, which have surrounded the bird. Captive birds, when they died, were usually stuffed, but the art of taxidermy was unsophisticated in those days and the stuffed birds bore only an approximate resemblance to what they had looked like in life. Moreover, the means for preservation were few: the animals were simply padded out with straw or tobacco, and within a relatively short time began to decay, which in turn led to more distortions of shape.

But there is another question: if the dodo was so horrible to eat, can it really have been hunted to death? Although there was probably never a huge basic population, it would have had to number at least tens of thousands if in-breeding was to be avoided, and all the sailors' reports suggest that it existed in large numbers. Wanton

killing of the birds for sport does not seem to have occurred to any serious extent, and Mauritius was not permanently or densely populated by humans until long after the dodo had vanished. So why did it die out so fast? Although it vanished relatively recently, we know very little about it – less than we know about the dinosaurs.

Everything now seems to suggest that it was, if anything, a victim of its own success, but the fundamental questions that led to this conclusion – what was it exactly? How did it get to Mauritius? Why did it have that unusual look? – did not find answers until very recent times. This is despite the popular appeal the bird has enjoyed since its discovery, and despite the intense interest it excited in those scientists and natural historians who followed Darwin in the nineteenth century. It was not until the year 2000 that DNA expert Dr Beth Shapiro took a sample from the bill of a dodo preserved for hundreds of years in the collection of the Oxford University Museum of Natural History. Possibly from the selfsame dodo that Hamon Lestrange had seen in London in 1638, though there is a record in the diary of one Thomas Crosfield that in 1634 a Mr Gosling bestowed a 'Dodar (a blacke Indian bird)' on the Oxford anatomy school, and Elias Ashmole himself may have got his example from there.

From the DNA sample taken from the bill, Shapiro was able to confirm that the dodo was a kind of pigeon. The dodo's small, flying ancestor must have made its evolutionary journey, perhaps by chance, blown by strong winds, and possibly via a chain of volcanic islands that have long since vanished beneath the sea. The

A reconstruction of the dodo in the Oxford University Museum of Natural History.

migration may have begun as much as 30 million years ago, with the bird arriving on Mauritius some 20 million years later. Ordinary pigeons are good fliers, but also well adapted to life on the ground, as any modern city dweller knows. Island-hopping would have been no problem for them but, having found a desirable environment, they would tend to stay put. Remains have been discovered on Rodrigues Island of another large flightless pigeon, the solitaire, and reports persist – though no hard evidence has come to light – of a 'white dodo' that inhabited Réunion.

How and why the dodo evolved as it did remained a mystery. Certainly its earliest ancestors must have arrived in a flock of at least thirty birds if in-breeding was to be avoided. When a bird becomes flightless, there is an immediate reduction of the muscles of the wing and pectoral girdle. The keel is reduced or lost, along with the flight muscles. Flightlessness can develop very fast – some scientists believe that it can happen within 10,000 years.

Because so few relics of the birds remained, scientists had very little to go on. The bird picked up and preserved by the eclectic collector Elias Ashmole in the seventeenth century was thrown away, with the exception of the skull, beak and one foot, by a spring-cleaning curator at the Ashmolean in 1755. Luckily these parts were preserved by an enlightened new curator, William Huddersfield, in the same year. Huddersfield had plaster casts made of them for the benefit of other museums, but after the mid-eighteenth century the trail of the dodo went so cold that by the 1800s naturalists began to wonder if the dodo had ever existed at all. One scientist, writing in 1801, described it dubiously as a 'feathered tortoise'.

However, just over half a century later, in 1865, at a remote place on Mauritius called Mare aux Songes (near Mahébourg in the southeast of the island), a local schoolteacher and amateur naturalist called George Clark came upon the first fossil dodo skeleton. Mare aux Songes is an old stream bed and the earth had eroded to expose a jumbled mass of bones, which luckily Clark was able to make sense of. His discovery turned him into a national hero, and today his picture can be found on Mauritian postage stamps. For the first time the true nature of the dodo could be researched. The skeletons – for several were discovered at the site – reveal that the dodo was indeed

a big bird, standing about 75 centimetres high. It had a long, sinuous neck. Its wings were undeveloped, about the size of a chicken's, confirming the dodo's inability to fly. Like other flightless birds, it had a small breast but strong, thick legs. Its head in proportion to its body was very large indeed – bigger than that of any existing birds – and it had a long, hooked bill.

Ornithologist and palaeobiologist Julian Hume, of London's Natural History Museum, has made a special study of the dodo and his researches have brought about a much closer understanding of what the bird was really like. On Mauritius today, where the dodo naturally also appears on national postage stamps, there is a monument at an airfield attesting that this is the site of 'the last abode of the dodo'. In fact the site it refers to is not Iversen's islet but the Mare aux Songes where the first fossil dodo remains were discovered, and it is just over a kilometre east of the monument. The site is now in the middle of a private sugar cane estate, and is situated where a rubbish dump once was. The flat area that forms a depression at the base of a range of low hills is still a pretty well-kept secret. In the Mauritian summer it is bone dry. In winter the rain fills it with around a metre of water. Since Clark's time, fifty-odd fossil dodos have been discovered here (it was thoroughly investigated in 1889 under the surveillance of Théodore Sauzier), and Julian Hume believes that the site is now almost fully excavated, though there may be other similar sites on the island where further remains may be discovered. What brought so many dodos to die in one place is not known: it may be the result of an accumulation of remains over a long period, or a disaster in which a flock of dodos wandered into marshland and became mired. Alternatively, the bodies may have been driven here by a cyclone. There is no evidence that this was a nesting site.

The keratin tip of the beak was shed and re-grown annually, and was used for eating and fighting in the jungle undergrowth world in which the dodo lived. In males a large bill in the breeding season might have been another display feature. Annual ridges of keratin may also prove to be a way of estimating a bird's age. The beak was powerful and would have equipped the dodo for a very wide diet. As for its environment, the forest of ebony trees that existed then would have been perfect. Ebony trees are tall and slender, so the forest floor would have been clear. Other trees, such as the latania and the pandanus, had plentiful fruit that grew close to ground level. These fruits would be accessible to tortoises and geckos, for example, in the interest of the plant's propagation, and were probably the dodo's staple diet. To feed such a large body, dodos must have spent a considerable part of their waking time eating, active generally at dawn and dusk, and would have scooped up their food quickly the moment they saw it, before others came near.

The dodo's diet cannot, however, have consisted merely of fruit. Julian Hume has pointed out that Mauritius is a volcanic island, with acidic soil, which means that calcium would have been in short supply. The dodos would have needed calcium for their eggshells, and so supplemented their diet with snails, crabs and coral – all of which were in plentiful supply on the island. The many early references to large gizzard stones suggest that the dodos needed them to grind down tough substances like crab or snail shell. Being big also meant that they had room for a long gut, which aided digestion – particularly of leaves and similar vegetation which was difficult to absorb.

Through these discoveries, a picture begins to grow of a highly adapted, sophisticated creature, able to take advantage of all the resources of its island home, and enjoying the fact that there were no predators. The only mammals native to Mauritius in recent times appear to have been fruit bats and insectivorous bats. In the sea were dugongs (sea-cows). Reptiles included various geckos, lizards and snakes, as well as two species of giant tortoise, both of which followed the dodo into extinction within half a century. Mankind introduced many mammal species to the island: cats, dogs, mongooses, pigs, deer, cattle, goats, horses, mice, rats, hares, rabbits, house shrews, tenrecs and monkeys among them. Rats, of course, would not have been the only animals in this group to breed fast, but they would have posed a threat to the dodo's eggs.

Before man's arrival, though, Mauritius was a paradise for the dodo. Because there were no predators and because there was plentiful food at ground level, the need to be able to fly became redundant. The action of flying requires enormous energy and muscle development – in some pigeon species the pectoral muscles account for a third of the total body-weight. The development of such muscles requires expensive nutrition. On the other hand, if all you need to do is walk, you can devote that nutrition to building up your size, which carried with it the advantages of slower metabolism, greater energy reserves and a longer life. Dodos may well have been capable of living for forty years. Dodos had, for birds, huge bodies, huge pelvic bones and a huge skull. But extreme specialization for its ground-dwelling existence made it vulnerable. Any changes to its environment would pose a threat to its existence.

Another myth that can be exploded is that the birds were docile. It is a characteristic of many species of pigeon that males have multiple fractures to their bones – the result, ornithologists believe, of fighting over mates. Like other pigeons, when it came to mating, which probably occurred at the turn of the year, the dodo male was very visual, noisy and violent. Big and without enemies, dodos would not have needed to flock together, except perhaps in lean times, in order to help each other find food – though this is conjectural. They would have probably only been really sociable in the rainy season, which was also the mating season; but then it is fair to assume that competition would have been fierce, and that the males would have been highly territorial.

Following the behaviour of other pigeons it is probable that dodo males would have staked out a nesting territory and gathered the raw materials for their ground-level nests. A good nesting site and the best available materials – palm fronds, lush grass – would help attract a mate, but a mating dance, performed in a specially prepared arena, was also part of the ritual. Interestingly, there seems to have been relatively little difference, except in size, between male and female dodos. The difference between the sexes in the dodo's cousin, the solitaire on Rodrigues Island, was very great. Solitaires may have organized themselves sexually into harem groups – a feature typical of greater sexual dimorphism.

Once a pair was formed, the female used the material provided by her mate to build a nest, and in it she would lay a single egg – white in colour, like all pigeons' eggs. The male would stay with her: pigeons exhibit strong pair-bonding, and it is not unlikely that dodos mated at least for the entire period of bringing up the single chick. The bigger the bird, the longer the process of development, so that it is possible that the chick would take nine months to reach maturity. This is one of the reasons why it is also likely that dodos bred only once every two years – so their reproduction was slow. Once weaned, the young dodo was apparently driven away to establish its own territory – the early French naturalist Legaut speaks of crèches of young in the upland forests, but no bones of young dodos have survived. Because of their size, individual dodos would have needed a respectably sized feeding territory.

George Clark made his discovery of the dodo fossil remains five years after the publication of Charles Darwin's *The Origin of Species*. His concept of the survival of the fittest and of evolution through natural selection caused uproar in Europe and America, and had no small influence on the interpretation both of the dodo's appearance and its fate. The zoologist Richard Owen, initially a disciple of Darwin, reconstructed the fossil skeleton found by Clark and fleshed it out in accordance with drawings and paintings from the seventeenth century, but of the bird in captivity – notably a picture by Roelandt Savery executed in about 1626. Owen was in a hurry to get his findings out – rival scientists at Cambridge University, Alfred and Edward Newton, were hot on his trail.

Owen's reconstruction seemed to fit the bill perfectly: a fat bird, slow and easy to kill – an obvious victim of natural selection. But how thoroughly had Owen done his homework? A massive study of the dodo by the ornithologists Strickland and Melville, published in 1848, does not appear to question Owen's reconstruction.

Biomechanics expert Andrew Kitchener is sceptical of the view that extinction due to developmental inadequacy is a tenable concept. He decided to build a model dodo which could realistically operate on the bone structure of Clark's skeleton, and his analysis of the weight those bones could actually support quickly scotched the old view of an overweight bird. If the dodo had been as fat as Owen and some of the earlier Dutch commentators had supposed, it would barely have been able to walk, given the skeleton it was equipped with. Dutch accounts had given the dodo's weight at something approaching 25 kilograms. Kitchener's view is that a male dodo

would have weighed in at around 17 kilograms – still heavy, but not improbably so.

There is, however, an explanation why some Dutch sailors, even on Mauritius, reported a fat bird. Many creatures store up fat in the rainy season to give themselves fuel for the dry months ahead. On Mauritius this cycle of seasons exists, and what is more the native animals have to be able to withstand tropical cyclones, which leave little standing in their wake, though in Mauritius nowadays at least cyclones are rare and seldom catastrophic. In any case, they would have the effect of bringing fruit down to the ground, and washing other food in from the sea. It is likely however that the omnivorous dodos would have gorged themselves in the wet season as an annual insurance policy against lean times ahead. Having said that, a dodo as fat as that proposed by Owen could scarcely have got around to feed itself, let alone mate or fight. Kitchener points out that dodos could not have got as fat as the Dutch maintained – to have done so they would have had to swing their body weight by 70 per cent every year, something that no animal on Earth does.

We now have a picture of an intelligent and well-adapted, ener-getic and competitive bird, which also had the advantage of being hard to cook and almost inedible – oily and cloying – when cooked – or half-cooked, as the Dutch seem never to have brought a tender dodo to their table. It seems unlikely that the dodo was hunted to death. So what brought about its dramatically sudden disappearance?

Dutch archaeologist Pieter Floore has investigated the site of Fort Hendrik, the permanent base established by the Dutch in 1638 on Mauritius as a staging post for their ships on the spice route to the East Indies. He and his fellow archaeologists have been able to find the bones of turtles, sea-cows and small birds among the kitchen waste at the fort, but no trace of dodo leftovers. It is interesting, however, that when Volkert Iversen discovered that last outpost of dodo survivors, he did so on an islet – possibly the ële aux Cerfs or the ële d'Ambre – north of the main island, as far from Fort Hendrik as they could possi-bly be. One explanation for this may be destruction of habitat: Fort Hendrik had a second function as a collecting point for the ebony that was being logged radially into the forests from it. Ebony was used for piano keys and for fine furniture, and commanded fabulous sums.

However, the facilities for logging were limited. Fort Hendrik never had a greater population than 300 people. Weather conditions on Mauritius – hot and wet – made it unpopular; there were mutinies, and the place never really flourished. Logging activities in proportion to the whole island were small, and in any case the hinterland was protected from man's incursions by dense, impassable rain forest. So destruction of the ebony forest cannot be held to be the reason for the dodo's extinction.

However, along with the bones of the native animals discovered in the middens of Fort Hendrik, Pieter Floore has found the remains of cattle, deer, goats, chickens and pigs. There are too many of them for it to be likely that they were brought to Mauritius as carcasses. The implication is that they were brought alive on ships, and ultimately allowed to breed in the forests to provide a limitless and easy food supply. In the forests they could be hunted, providing the colonists with a much-needed source of recreation.

But their arrival was a disaster for the dodo. The rich under-growth that made Mauritius such an ideal place for the dodo was also perfectly suited to pigs and, of all the animals introduced by the Dutch, it is the pig which is the most likely to have sealed the dodos' fate. For maybe millions of years the dodo had been free from pred-ators and it had no rivals for its food supply. As it had nothing to fear, it laid its eggs on the ground and incubated them there. There was only a single egg per family, laid every two years, and the chick would take up to a year to reach maturity. That in itself was not a problem; before the advent of man there was no reason why nearly all the hatchlings should not survive. A greater birth-rate might have led to overpopulation, so the pace of the dodos' reproduction was logical.

But the pig changed all that. It ate the ground-level fruit, it dug up roots and tubers, it probably trampled over nest sites and broke the eggs. It bred with dizzying speed, too – a visitor to the island in 1709 describes a pig-hunt in which 1,500 animals were killed in the course of a single afternoon. With fewer and fewer young, the surviv-ing population of dodos had an increasingly older average age. The spread of pigs drove them farther and farther away from Fort Hendrik, where the pigs had started out from, but rain forests were no hindrance to the pigs, which were hardy beasts and adapted easily to their luxurious, food-rich circumstances. And all this happened far too fast for the dodo to adapt.

The surviving community discovered by Iversen would have been under considerable stress. Underpopulation would have made it harder to find mates, and competition for the mates there were would have been correspondingly more desperate and aggressive. But most would not have bred at all: this would have been an ageing, declining population. There may well have been enough food – the dodos reaching the islet by way of a sand-bar exposed at low tide, and in any event pigs do not seem to have reached this last outpost by the time Iversen arrived – but there were no longer enough dodos: they had become biologically extinct. With such a slow cycle of reproduction, the few squabs that were born could never replace the numbers that had gone. The dodo was spiralling out of existence, undermined by its own superb but sadly overspecialized adaptation to its environment.

In 1693 the French explorer and naturalist François Legaut reached Iversen's islet but did not find a single dodo. The bird – which has become a symbol for extinction, part comic, part tragic, and oddly lovable – had vanished from the Earth well within a century of its first encounter with man.

BACKPLATES

The backgrounds for the computer-generated sections of each episode of *Extinct* were shot over spring 2001 in Oregon, California, Mauritius, Tasmania and Iceland.

They're all very different locations, but the way in which the backplates were shot is fundamentally the same. In theory, shooting a backplate is simple. Just find the best, most dramatic vista, mount the camera on a tripod, turn it on and stand back while it records the scene for a minute or so. What you end up with is a tableau into which the computer animal can be placed so it appears as if it's moving in a real environment.

In reality, however, it isn't quite as easy as that. Backplates have to be shot in a certain way in order to make the animation convincing. The hardest thing to reproduce is the impact a creature has on its surroundings as it moves. A real animal constantly interacts with the world around it. As it walks it kicks up dust, squashes grass and breaks the twigs it steps on, while the light throws shadows on to the animal's fur or feathers. An animal moving through water disturbs it in a complex and unpredictable way, and it also has a reflection. The ripples caused by throwing a pebble into a still pond can be created by a computer, with difficulty, but imagine the splash a penguin (or a great auk) makes diving into the ocean as it crashes against the shore.

In short, a myriad complex interactions between surfaces, light and other elements occur when even the smallest movement takes place, and all of these interactions have to be replicated. The only way to do this is to choreograph the impact an animal would have on its surroundings if it was actually there.

For instance, if we were to make an animal walk through vegetation, brushing aside plants as it did so, we'd have to physically do that ourselves, in one of several different ways. One method is to get someone to mimic the creature's movement, then paint them out of the shot frame by frame, leaving the movement of the vegetation intact, before replacing the human with the computer-generated

On location in Mauritius, a toy dodo stands in as a position marker in a nest.

Filming the dodo
reconstruction scenes:
actors in period costume
as Dutch pirates (above);
making dodo footprints in
the sand (right).

The crew on location on the beach at Ilôt Sancho, Mauritius.

animal model. This is a very time-consuming process and there's no guarantee it will always work sufficiently well.

A second solution is to attach invisible nylon threads to the individual plants, and use them to rustle the vegetation so it appears to move with the projected walk of the animated creature. This is a much simpler option, and requires less post-production work, but it's difficult to mimic the impact a large animal like a mammoth would have on its surroundings. We had to use other techniques to reproduce the impact of a heavy beast like a mammoth, and that's where blue-screen technology came into play.

Filming against blue screens to produce what is known as a chromakey has been in use in film and television for several decades. When a shot is 'keyed', it's layered on to another scene and the solid blue background colour allows the elements to combine seamlessly. On *Extinct*, blue screens proved indispensable for our special effects. Very often, foreground vegetation, behind which the animal appears

Wallabies in the Truwunna
Wildlife Park, used for the
Tasmanian tiger shoots.

to move, has been shot as a separate image against a blue screen. A plant frond is held by a clamp in front of the blue screen and filmed. This can then be added to the final backplate and animation to give a sense of depth, creating the impression that the animal is within the landscape, not simply placed on top.

We also used lots of other techniques and tricks to enhance the reality of a scene. Some are pretty basic, like using a smoke machine to add a wisp of early morning mist in a forest, or a wind machine to move the vegetation. But in certain backplates, the entire landscape is actually a composite image made up of different elements. As well as foreground interest, mountain ranges may have been added in on the horizon, for example. A clear blue sky can make an ideal natural blue screen, weather permitting, against which birds or any object can be placed.

This illusion can be much harder to achieve when an animal has to interact with water, and we tended to try and avoid it on *Extinct*, but there was one episode where it had to be done – the great auk. As a sea bird, the auk constantly interacted with water: diving, swimming and bobbing on the surface, everything it did involved water. Surprisingly, underwater shots aren't actually a problem, as brilliantly realistic undersea environments can be generated on a computer, eliminating the need to undertake complex and potentially hazardous underwater shoots. Above water, there's no problem either, and the backplates can be put together like any other. The problem comes when the animal has to move from one environment to the other, for example diving into the sea and splashing through the waves. Nylon cords can't be fixed to the water, and blue screens can't be set up on the sea, so the solution we opted for was to drop weights into the waves ourselves, on location, film the splashes they made and then digitally remove the weights from the shot, replacing them with computer-animated great auks.

Reconstructing the moment when farmer Wilf Batty shot dead the last Tasmanian tiger recorded in the wild.

With the last backplates shot, the computer animation that would bring our creatures back to life could begin.

GREAT AUK

Pinguinus impennis

The Icelandic summer is brief. You could say that June is spring, July is summer and August is autumn; during these months the days rapidly lengthen, and for the middle weeks of the period there is daylight twenty-four hours a day. But after that the days shorten just as rapidly, and winter, when it comes, is long and dark.

On 3 June 1844 a group of fishermen set out from Keflavik for the island of Eldey – Fire Island – just off the extreme south-western tip of Iceland. Eldey was not a welcoming place. Sailors feared its jagged rocks and the raging seas surrounding it, but the fishermen knew that there might still be a few great auks here – the remnants of a once-thriving colony that had been virtually annihilated by hunting. Now, with so few left, the value of the birds to collectors had risen astronomically, so the men were prepared to risk the perils of landing on Eldey. The expedition was led by Vilhjälmar Hakonarsson, and it managed to land three men on the island.

The (anglicized) name that the Icelanders gave the bird was 'garefowl'. Garefowl means spear-billed bird, and probably comes from Old Norse via the Icelandic *geirfugl*. The large, 70-centimetre-long bird had been known to man for millennia, but only since the mid-eighteenth century had it been threatened with destruction, and no one had really bothered to study its habits. But even in 1844, when the great auk's rarity was acknowledged, no one seriously believed that it was on the brink of extinction. Certainly the fishermen who landed on Eldey that day early in June had no idea what a fateful role they were about to play.

If anything, they might have been disappointed, for they tracked down only one pair of birds, which ran from them, not even attempting to defend themselves with their razor-sharp beaks. The fishermen soon caught and strangled them, taking care not to damage their feathers or break their skin, for they were working on commission – a dealer in Reykjavik had a purchaser already lined up. Unluckily, the one egg they discovered, which the birds had been taking turns to incubate, was broken, so they left it where it was. A nineteenth-century account written by the ornithological historian Symington Grieve tells the story:

As the men clambered up they saw two garefowl sitting among numberless other rock-birds (murres and razorbills) and at once gave chase. The garefowl showed not the slightest disposition to repel the invaders, but immediately ran along under the high cliff, their heads erect, their wings somewhat extended. They uttered no cry of alarm, and moved, with their short steps, about as quickly as a man could walk. Jón Brandsson [one of the three fishermen involved in the actual hunt], with outstretched arms, drove one into a corner, where he soon had it fast. Sigurör Islefsson and Ketil Ketilsson pursued the second, and the former seized it close to the edge of the rock, here risen to a precipice some fathoms high, the water being directly below it. Ketil then returned to the sloping shelf whence the birds had started, and saw an egg lying on the lava slab, which he knew to be a garefowl's. He took it up but finding it broken put it down again. Whether there was another egg is uncertain. All this took much less time than it takes to tell.

Hakonarsson returned to Eldey in 1846 and again as late as 1860 in the hope of finding more great auks, but without success.

The birds they killed were probably not the last individuals alive – a reliable sighting was reported in 1852 – but these were almost certainly the very last great auks killed by man. All that was left of the great auk, the only flightless seabird ever to live in the northern hemisphere in historical times, were a few stuffed examples scattered through museums across the world, some eggs, and a handful of skeletons. It remained a mystery to science, and the greatest tragedy was that the bird need not have become extinct at all: man's fecklessness alone is the reason for its demise. In fact, the great auk is the only member of its family, the Alcidae, to become extinct during historical times.

Before it became the object of remorseless hunting, the great auk's habitat had been widespread. Living most of its life at sea and ranging right across the north Atlantic, it came ashore only to breed, and because it was ungainly and cumbersome on land it sought out flat, low-lying islands on which it was easy to beach, and which provided rich inshore feeding. It may have roosted at sea, apart from the breeding season, but its behaviour generally while at sea is unknown, as it was hunted to death before it could be properly observed.

Probably in reaction to its predation by man, the islands the great auk needed had to be small, uninhabited by other animals – Arctic wolves and polar bears were the bird's natural enemies, apart from man – and far enough away from the mainland for the birds to be able to breed securely. It has been argued that great auks would not have chosen to breed too far north – that is, beyond the Arctic Circle – since there they would have been easy prey for such creatures. If auks were ever threatened by predatory gulls such as skuas, it may be that they were big enough and well enough armed to defend their eggs or chicks successfully.

Accounts dating from relatively modern times tell us that great auks stood almost upright on land, as penguins do. In summer their plumage was a blackish brown on all the upper surfaces, with narrow white tips on the secondary wing plumage and large white spots on the face between the bill and the eye. Although their throats and chins were dark, their chests and bellies were all white. Bill and feet were black, and the webbed toes ended in short claws. The bill had a series of vertical grooves, between six and twelve, on both the upper and lower mandible towards the tip, the eye was a dark, chestnut brown, almost black, and the inside of the mouth probably an orange-yellow. In winter the white patches in front of the eyes reduced to a narrow white line, and the dark throat feathers gave way to white. Early naturalists working when the auk was still alive were able to gather further information, such as weight, estimated at about 4 or 5 kilograms, allowing for a slight difference in body size between males and females. The wings, used for propulsion through the water (the webbed feet were used as a rudder) were very muscular but relatively small: only about 17 centimetres long.

One secure breeding place was Funk Island, a tiny rocky islet just under a kilometre long and less than half a kilometre wide, which lies out in the Atlantic off the north-east coast of Newfoundland. North-west and south-west shores slope gently into the sea, and are where the great auks came and went for the most part. Generally bleak and lashed by storms, Funk is not an easy place for a boat to put in and, if the weather is really bad, it is impossible; but the ecologist William

Watercolour painting of the Auk by William MacGillivray, c.1831.

Montevecchi has been going there annually for several years in an attempt to piece together more information on the great auk than was ever collected when it was alive.

Low-lying islands may not have been an absolute factor limiting breeding sites. The rockhopper penguin can climb 100-metre-high cliffs in search of a suitable nesting site. This in turn suggests that the great auk may have been capable when pressed of colonizing islands hitherto deemed inaccessible to it, such as the Bird Rocks (the Magdalen Islands), Eldey and parts of St Kilda.

Funk Island is the last landfall heading east from North America until you reach the English coast 3,000 miles away. It is hammered by the Atlantic, stinks of ammonia – the funk that gives the island its name, and which derives from the uric acid in bird excrement – and seems the most inhospitable place on earth. But it is a breeding-ground for guillemots and razor-billed auks, and until the end of the eighteenth century it was home to the largest colony of great auks in the world. It was also the scene of one of the most profligate acts of destruction to nature man has perpetrated.

The presence of the great auk here is confirmed by examining the thin carpet of soil that supports hardy grasses and a few stubborn shrubs that can withstand the harsh conditions, and survive in the salt-laden air. The soil itself owes its existence to the millions of birds that have bred here over centuries, if not millennia: it is a dense compost, made up of decomposed excrement and corpses.

Funk Island lies near the great cod fisheries of the Grand Banks. Until recent times these seas teemed with fish, providing an unexampled supply of food to the marine birds of the region, and it would appear that the Funk great auks were somewhat larger than their relatives elsewhere. But the Grand Banks were already known to the intrepid seafarers of England, Spain and Portugal by the early sixteenth century. John Cabot, under commission from King Henry VII, touched on the coasts of what would become Nova Scotia and Newfoundland on his second expedition in 1498, convinced that he had reached Asia. Funk Island first appeared on a chart in 1501, where the cartographer, Pedro Renel, who had accompanied Gaspar Côrte-Real in his explorations to the New World, calls it the Y dos Aves – 'Island of Birds' – an indication that the auks were already there, as they no doubt had been since archaic times. Soon afterwards, a total of around 350 ships a year from France, England, Spain and Portugal were regularly making the voyage out to the Grand Banks to fish.

Before these relatively modern explorers, Vikings may also have reached Funk – they certainly reached Newfoundland – and there is evidence to suggest that local aboriginal human populations were hunting the great auk at least 1,100 years ago and even earlier. As long ago as AD 500 Beothuk Indians canoed to Funk Island to collect great auk eggs, according to Montevecchi and Tuck, writing in 1987. The eggs, being three times the size of those of the murre, would have been well worth the effort of the rough crossing.

As the early navigators crossed the Atlantic westwards towards the edge of the known world, they reached Funk Island and discovered a strange, flightless seabird, which they called a 'penguin', a reference either to its white head (the great auk had white markings on the side of its head in front of its eyes in its summer plumage, and the Welsh for 'white head' is *pen gwyn*), or to the fact that its wings looked pinioned (pen-winged), or again because the bird was fat (Latin *pinguis*). The word 'auk', from the Swedish *alka*, did not become generally current until the late seventeenth century, and in French the name for an auk is still *grand pingouin*. In Spanish it is *pinguino grande*.

At the time, the birds now called penguins, which live in Antarctica, were unknown. The great auk, which shares some characteristics with the penguin, and indeed looks like a penguin, was the first to carry that name, and later bequeathed it to the southern hemisphere birds, which are not related.

Contemporary accounts spoke of them thus: 'Some of these birds were as large as geese, being black and white with a beak like a crow's. They are always on the water, not being able to fly in the air, inasmuch as they have only small wings about the size of half one's hand, with which however they move as quickly along the water as the other birds fly through the air.' It was impossible to catch an auk in the water, even when pursuing it in a six-oared boat. On land, it was easy to pick them up, if you could avoid the beak. There is a report of a fisherman off the coast of Greenland, at Gunnbjörn Rocks, filling his boat with garefowl in a morning. Funk Island acquired the name Isla de Pitigoen (Penguin Island). A number of men could slaughter enough great auks to fill a boat within half an hour, and there are reports from the first half of the sixteenth century of four or five tons of the birds being salted in barrels.

By 1536, Newfoundland's fisheries and wealth of food sources were so well known in Europe that a group of London merchants chartered and equipped two ships to take 'thirty gentlemen to view its wonders and visit the island of the penguins'. They enjoyed 'a great feast of great auks' as well. Nearly a century later, Captain Richard Whitbourne, in his *Discourse and Discovery of Newfoundland* (1629), could write regarding the great auk: 'These Penguins are as big as geese... and they multiply so infinitely upon a certain flat island that men drive them from hence upon a board, into their boat by the hundred at that time, as if God had made the innocencie of so poor a creature to become such an admirable instrument for the sustentation of man.' Some modern scientists, however, have doubted whether great auks could indeed have been marshalled along a plank on to a waiting vessel.

William Montevecchi has been able to confirm the truth of the old accounts of the great auk by comparing the observations in them with bone samples found on Funk Island. They show that the great auk, unlike the present residents of the island, was indeed flightless, though it was closely related to guillemots and razorbills. What is astonishing, however, is the number of bones that remain, and the existence of the soil itself attests to a vast population of great auks breeding here over many centuries. During the breeding season in June, as many as 100,000 pairs would have gathered on the island. The pairs, which probably bonded for life, produced one egg a year, laid on the bare rock. Its great numbers testify to the success of its highly specialized adaptation.

Skeletons alone can tell little of the great auk's behaviour, of course, though it can be established that, clumsy as they were on land, they were perfectly designed for a life in and under the water, which is where they spent 90 per cent of their time. To find out more requires the information supplied by remains collected in natural history museums. The skinless corpses of the last great auks killed on Eldey ended up in the Royal Museum, Copenhagen, preserved in spirits. The fate of the skins, also originally sent to Denmark, is unknown. Before photography, despite often quite accurate drawings by such artists as Edward Lear, the only way for naturalists not in the field to gain a true impression of what an animal looked like was through stuffed or pickled specimens. If an animal became extinct before photography, that is still the only material naturalists have to go on.

From stuffed examples of the great auk, it is possible to deduce that its closest living relative is in fact the razorbill, which is not only

able to fly, but is only a fifth the size of the great auk. The fact that the great auk is similar in appearance to the penguin is an example of parallel evolution. The forces of natural selection acted on a bird entirely different from the penguin to produce something with the same body form – designed to perform similar functions. The great auk, however, was larger than most penguins.

As soon as the great auk opted, as it were, for the sea rather than the air, it developed over millennia into one of the most efficient divers ever.

The bigger and heavier a bird becomes, the easier it is for it to dive. Large size means a relatively slower metabolic rate. The more slowly it uses up air, the longer it is able to stay underwater. The great auk developed a thick layer of subcutaneous fat to keep it insulated and warm, and its body form was perfectly hydrodynamic. It had very short, muscular wings, which were perfect for providing propulsion through so dense a medium as water, and there were further refinements: every single feather had its own special muscle, enabling the great auk to flatten its feathers close to its body, trapping as little air as possible, thus making it less buoyant and increasing diving efficiency. In the sea, the bird was a superb swimmer, extremely fast and agile, flying underwater. From early accounts and the analysis of remains, it is possible to create a picture of what they ate, and what, having a great gape, they could swallow whole, since it was better to

A stuffed great auk specimen.

do that at sea than waste energy on piecemeal eating and digestion. Off the coasts of Greenland it appears that they ate shorthorn sculpins and lumpsuckers. On Funk Island remains have been found to contain Atlantic menhaden up to 190 millimetres long, shad, capelin, stickleback and striped bass, among others.

How the great auk managed to dive as deep as it did – down to 100 metres and more – and hold its breath for anything up to fifteen minutes are questions to which scientists are still seeking answers. How diving birds avoid the bends, the decompression malady that can affect human divers, is not certain. Their lungs may not collapse during a dive, and since atmospheric air is nearly 80 per cent nitrogen, their air sacs would serve as a very good nitrogen store. Laboratory experiments with diving Adélie and gentoo penguins, both medium-sized species, indicate that a gas exchange continued throughout the dive: the oxygen in the lungs became progressively depleted, and the nitrogen in the blood rose to levels that reached the borderline – but did not cross it – for decompression sickness. It is believed that these birds avoid decompression sickness by keeping their dives short, and in the wild it has been observed that most penguins do not dive deeply enough or for long enough periods to incur a risk from excess nitrogen during any one dive.

The reason for developing these abilities was to hunt fish, but that raises another question – how did the great auk find its prey? Diving birds must evolve to cope well in two very different environments: on dry land and underwater. They rely on the ocean environment for food and travel, yet they must return to the surface to breathe and to the land to breed. Water and air refract light differently, so air-adapted eyes cannot focus properly underwater, and terrestrial animals become 'long-sighted'. Seawater also absorbs and scatters sunlight, so the deeper we go, the darker it gets, until at depths of 100–150 metres, or less in murky waters, no light can penetrate and the ocean is effectively pitch black.

We can't know just how specialized the great auk's eyes were, but by examining birds that inhabit similar environments, such as penguins, we can speculate on the kind of sensory adaptations it may have had. Scientists have discovered that penguins are able to see normally in both air and water, thanks to strong eye muscles that contract and warp the lens to compensate for the degree of light refraction. Penguins have large, flat eyes to maximize light intake, and they are particularly sensitive to blue, green and even near-uv light – an adaptation to the blue-green ocean world in which they live. Animals such as cats, dogs and fish have a reflective layer at the back of their eyes, called a tapetum, which reflects light back on to the retina, improving vision in poor light conditions. This adaptation has only been found in one nocturnal species of bird, however (the nightjar), so it's unlikely that the auk had a similar facility, although not impossible.

This mighty hunter was equipped with a prodigious weapon. The great auk had a large, narrow bill that it could use when threatened as a formidable dagger. The ornithological historian Jeremy Gaskell has unearthed contemporary accounts that give ample proof of the great auk's aggression and willingness to defend itself. Aggression within densely populated breeding colonies is likely, and early observers mention that if caught, the bird would 'bite everyone within reach of its powerful bill' (Audubon, c. 1840) or 'bite wildly all around' (Fabricius, 1808).

It is not insignificant that the last great auks hunted down on Eldey fled from their captors – isolated small groups of surviving great auks would experience great stress. Auks naturally lived in huge colonies, and densely packed together, not unlike the bluefooted booby colonies in the Galapagos today. In 1718 a surveyor of Penguin Island off Newfoundland reported that French colonists of Placentia told him that 'a Mann could not goe ashoar upon those Islands [sic] without Bootes, for otherwise they would spoil his Leggs, that they were Intirely covered withn [sic] those fowles, soe close that a mann could not put his foot between them'. It has been suggested that great auks' 'nests' were within one square metre of each other.

Over fifty years ago the American biologist Walter Allee put forward a new theory about why such a species might become extinct. He studied the consequences of low population density and realized that 'under-crowding' was an important factor in a given species' survival potential. Such birds as penguins and murres, which breed in vast numbers to protect themselves from predators, find it hard to breed when their numbers are seriously thinned. If this

GREAT AUK: VITAL STATISTICS

Pinguinus impennis (Linnaeus: *Alca impennis*)

Also known as: Magellan's goose; anglemager (Norwegian), after its cry, aangla, to tell fishermen to get their hooks (angler) ready; arponaz (old Basque, 'spearbill'), but also perhaps related to Inuit agparak; apponath (perhaps from the Beothuk Indian word for the bird); garefowl; gorfou; penguin.

Plumage: same for both sexes (adults). Summer: blackish brown above, with narrow white tips on secondary wing plumage and large white spots on face between bill and eye; white below, except for dark throat and chin. Bill and feet black; webbed toes ended in short claws. Winter: white patches on face reduced to narrow white line; dark throat now white. Chick: grey down.

Weight: about 4–5 kg.

Size: about 70 cm, beak to tail. Slight difference in sexes.

Egg: pyriform.

Incubation: about six weeks; chicks ready for seagoing within perhaps two days after hatching.

Breeding: around June on at least eight identified traditional breeding grounds – small flat rocky islets offshore and selected for inaccessibility to predators.

Sociability: highly social birds breeding in vast colonies during summer months.

Range: at sea for most of year and capable of travelling great distances.

Food: larger fish (beak had wide gape) – e.g. lumpsucker and shorthorn sculpin as well as striped bass, capelin and shad; and crustaceans.

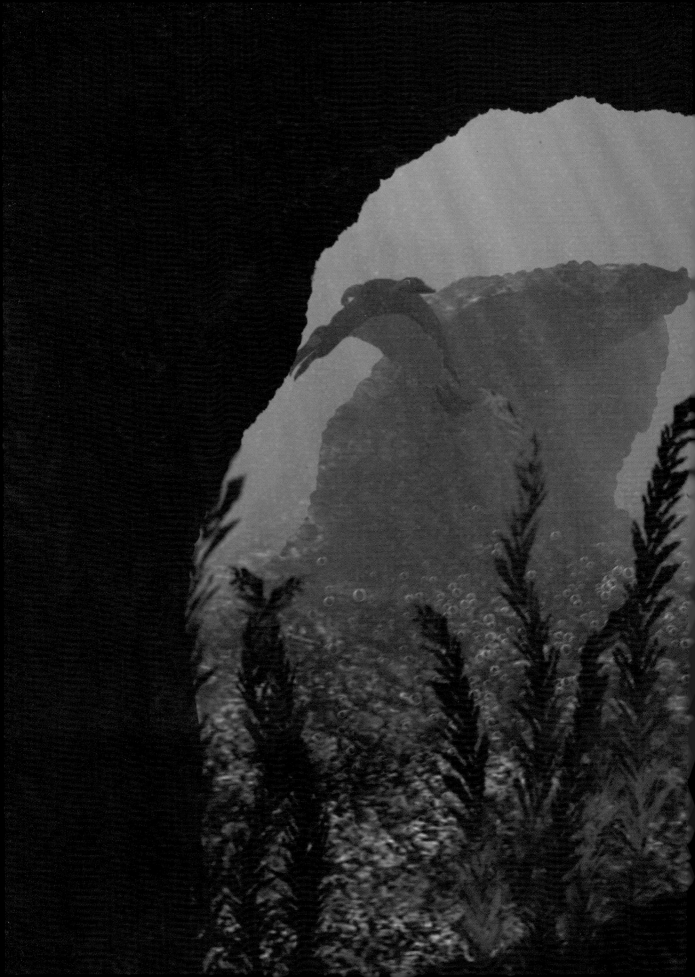

theory is correct, the last auks of Eldey had performed a near miracle by producing an egg at all.

So severe underpopulation would cause stress, demoralization and a tendency to flee rather than fight. But an account Gaskell discovered from Iceland in 1824, only twenty years before the raid on Eldey, tells how a young man was pecked so severely by a cornered great auk that in spite of the calfskin jacket he was wearing the bird was able to draw blood seriously.

Funk Island, for many thousands of years well out of the way of man and surrounded by fish-rich seas, must have been a paradise for the great auk. But it had to pay a price for its highly specialized adaptation to life in the water. On land it was profoundly vulnerable, and although it spent as little time there as possible, the times and places it did so were regular and predictable. Because it was ungainly on land, its early discoverers anthropomorphically attributed idiocy to it. Their prejudice is evident in their descriptions of it on land: it 'walked very awkwardly, often tumbling over' – it was 'stupid and tame'.

The great auk bred regularly in May to June (unfortunately for it, also a good time for sailing) depending on the location of its breeding colonies, which were always in the same places. The largest of these, Funk Island, was known to man by the early sixteenth century, as we have seen, and in the vastness of the Atlantic there were few alternative breeding grounds. As marine navigation became more efficient, as man grew bolder and more confident, and as fishing expeditions to the North Atlantic became an increasingly regular occurrence, the great auk colony provided a welcome source of food, and it would not be long before over-exploitation sealed the birds' fate. Even so, as late as 1728 the mariner John Seller could write: 'you may know when you are upon the Bank [the Grand Banks]... by the great quantities of fowls... none are to be minded [noticed] so much as the Pengwins [sic], for these never go without the Bank, as the others do.'

It is likely that the great auk's breeding habits were similar to those observed in the birds that breed on Funk Island today. It is not known whether copulation took place at sea or on land, though the latter is far more likely. Mating would have been preceded by an elaborate courtship ritual, and the parental pair would have taken turns in brooding the egg, one sitting on it while the other went in search of food. Mutual preening and beak-tapping would have formed an important part of the bonding mechanism, and though great auks do not appear to have been very vocal, soft calling to one another is also probable. The only noises attributed to them are 'low groans' and 'croaks'. The colony was so huge that it was important that a returning partner could find its mate without undue difficulty.

As the egg was laid on bare rock, its shape was important. A great auk's elegantly patterned egg was wide at one end and narrow at the other – so-called ovate pyriform. The reason for this is taken to be that it would have prevented the egg from rolling away; rather, it would have rolled in a tight circle, though the shape also suggests a maximum use of available space – great auks nested shoulder-to-shoulder and the egg's shape meant that it could be brooded by a bird standing upright. The changeover of partner was painstaking and could result in accidents. It is possible that if an egg was lost, the breeding pair would try to reproduce within the season. No records exist at all concerning the newborn chick, but from the known weight of the egg, scientists have determined that the chick weighed about 240 grams at hatching.

Incubation may have taken about six weeks, with the newborn chicks ready to take to sea in a matter of a very few days – perhaps even as few as two; though it should be stressed that these estimates are pure conjecture. There are no accounts of chicks at colonies and no remains of chicks have come to light; and even the matter of how and on what they fed is still a mystery, since they would have been too small to eat fish whole as their parents did, and there is no evidence that great auks were capable of regurgitative feeding. It is possible that the young fed on zooplankton of some kind, and took a year to reach maturity. The only description of young great auks comes from an account of a raid on the auk colony at Geirfuglasker, a cluster of islets and skerries 40 kilometres south-west of Cape Rejkjanes, in south-west Iceland. It was made by a British privateer in August 1808, a time that must have coincided with hatching, as both unhatched eggs and young are described. The British sailors spent several days on the island, clubbing the birds to death for sport as much as for food. Other reports describe downy grey chicks at sea,

Illustration of a great auk
by W. Lewin, c1789.

Illustration of a great auk
by W. Lewin, c1789.

and chicks riding on their parents' backs. If the chicks were at sea within a very short time of being hatched, a long period of parental care suggests itself, but with very little documentary evidence and no specimens of young to go on, the most scientists can do is make educated guesses.

The period during which the auks were at their most vulnerable was during the six weeks of incubation. Given the fact that it was at this time that they were hunted, and that they produced only one chick per pair, it seems incredible that they survived as long as they did, despite their vast numbers. Indeed, if most of the eggs were removed during the denuding of Funk Island, it is probable that very few chicks were born, and so the population must have taken a relatively rapid nosedive. It can only be assumed that the sheer number of birds on Funk was enough to absorb the depredations for several generations. Certainly the great auk did not learn to move elsewhere – though, as we have seen, its choice of breeding ground was severely limited anyway.

William Montevecchi has discovered a series of circular walls and low earth mounds on Funk Island. These are the remains of the pens that men built there, into which the auks were herded before slaughter. There is also evidence of a large number of fires. Initially the birds were killed for food, and the eggs provided a welcome source of protein after a long Atlantic voyage. What they tasted like is not known, though a surprising number of accounts describe the flesh as palatable. In general, the oily flesh of a fish-eating bird is not pleasant, but the mariners landing on Funk would not have been fastidious.

The presence of the pens, however, points to a more sinister activity than simply culling a few birds for food. The pens represent what had become, by the eighteenth century, summer extermination camps. The birds were slaughtered wholesale for their beautiful back feathers, 'soft and smooth as black silk', which had become an indispensable fashion accessory for trimming hats and dresses in Europe, and their oil, which was a valued fuel. A contemporary account by the French fur trader Nicolas Denys, whose keen general observation is worth quoting in full, testifies to the richness of the supply of oil:

> *The great auk (pennegoin) is another bird, variegated in white and black. It does not fly. It has only two stumps of wings with which it beats upon the water to aid in flying or diving. It is claimed that if it dives even to the bottom to seek its prey upon the [Grand] Bank. It is found more than one hundred leagues from land, where, nevertheless, it comes to lay its eggs like the others. When they have had their young, they plunge into the water, and the young place themselves upon their backs and are carried like this as far as the Bank. There one sees some no larger than chickens, although they grow as large as geese. All those birds are [considered] good to eat by fishermen. As for myself I do not find them agreeable. They make fish oil. The fishermen collect them for this purpose. There are vessels which have made as much as ten or twelve puncheons [over 5,000 litres] of it.*

Kept in the pens until the process of rendering was ready to begin, the birds were then killed and plucked. Their bodies were rendered in cauldrons to produce the oil – but the most profligate use of them was that auks' bodies also fuelled the fires on which the cauldrons were placed. Their bodies, full of oil, burned very well, and as there was no wood available on Funk Island...

The process of industrial slaughter reached its height by the middle of the century. By 1785, an English gentleman adventurer and trapper called George Cartwright, who had been living in Labrador since 1770 and was then anchored in Shoal Cove off Fogo Island, wrote in his journal what may be one of the first ecologically conscious observations on record:

> A boat came in from Funk Island laden with birds, chiefly penguins [sic]... it has been customary of late years for crews of men to live all summer on... that island, for the sole purpose of killing birds for the sake of their feathers... the destruction which they have made is incredible... if a stop is not put to that practice the whole breed will be diminished to almost nothing, particularly the penguins: for this is now the only island they have left to breed upon.

Cartwright was writing of the St Lawrence vicinity islands. The authorities paid heed to what he said, and in fact ten years earlier the governing council of Newfoundland had already petitioned the British parliament to stop the wholesale slaughter of seabirds, and magistrates at St John's introduced harsh penalties, including flogging, for those caught killing birds for their feathers or eggs. But it was a hard matter to police and regulate. Taking birds for fish bait was still permissible, however, though how use could be determined by the authorities is a mystery. In 1793 there is a record that Chief Justice Reeves convicted three men from Greenspond for taking great auk eggs from Funk Island during the closed season. That there were men with not only an ecological conscience, but also a sense of practical impoverishment should the birds disappear, is shown by the Reverend L. A. Anspach, who in 1819 suggested that the usefulness of great auks as aids to navigation was in danger of being lost to overexploitation.

The year after George Cartwright wrote his observations in his journal, 1786, a proclamation was finally issued banning the killing on Funk. But by then it was already too late. The evidence still available today shows that the slaughter was conducted on a truly massive scale, akin to, and arguably worse than, the seal 'culls' of recent times that caused such international outrage.

By the beginning of the nineteenth century, the great auk was extinct on Funk Island, and the species was lost for ever in Canada. However, the bird had not died out everywhere. Very little was known about its range at the time it disappeared from the north-west Atlantic, but historical records show that a substantial auk settlement existed on the Geirfuglasker – Auk Rocks – a short distance further out from the mainland than Eldey, before Eldey was colonized, and it is possible that wherever suitable offshore islands existed in the northerly latitudes of the northern hemisphere, there may have been auk colonies in antiquity.

On the face of it, the auk should have been much safer on the Geirfuglasker than on Funk Island. The rocky islets, though not very far from the mainland, were surrounded by ferocious seas, and to the local people they were places of mystery and terror. Stories abound of fishermen shipwrecked here and never seen again, and Icelandic sagas tell of the fairies and goblins that inhabited the skerries. One man who had been stranded on the Geirfuglasker was said to have gone mad, jumped off a cliff, and turned into a whale. More prosaically, in 1628, twelve men were drowned attempting to reach the Geirfuglasker, and eleven years later two large boats from a fleet of four vanished on a similar mission.

Given the dangers both natural and supernatural, one might have thought that here at least the auk would be left in peace, but that was not the case (certainly those British privateers were not deterred in 1808). However, the Geirfuglasker suffered similar depredations to those at Funk Island during the eighteenth century, though these were due mainly to egg-collecting expeditions. When the colony seemed in danger of extinction, the great auks were left alone for a number of years, allowing the population to recover sufficiently to make egg-collecting worthwhile again – though this apparent act of conservation was more by accident than design: the birds were spared for a time because the seas became too wild and dangerous for the Icelanders to make a landing.

Jeremy Gaskell has discovered that as the great auk died out in Newfoundland, local records indicate that the bird's European population also started to dwindle. The suggestion is that the great auk, a prodigiously strong swimmer, may have been capable of migrating long distances across the sea. The range of the auk is now believed to have spread across the northern hemisphere in an arc across the North Atlantic from Canada to Greenland and Iceland, and across to Britain. However, great auk bones of considerable antiquity have been discovered over a far greater area: in Florida (dating from about 3,250 years ago), in New England and Labrador, as well as Greenland, and along the coasts of Norway, Denmark, the Netherlands, Brittany, the Bristol Channel, and even Gibraltar. These, however (and those in Florida), may date from a time when the ice age caused the climate in those latitudes to be much colder than it is today.

The oldest remains discovered, dating to around 75,000 years ago, were at Gibraltar. Late Stone Age cave paintings, dating from around 18,500 years ago and showing great auks, have been located near Marseilles, and remains have been found in Norwegian kitchen middens across a great expanse of time – from about 13,000 to about 2,000 years ago. It is also possible that the bones arrived at some of their destinations via traders. Certainly no great auks have been reported in southerly latitudes, even as wintering birds, in historic times. The great auk, however, seems to have belonged to a long line of large flightless seabirds that lived in the northern hemisphere. An ancient form, *Pinguinus alfrednewtoni*, is patchily recorded as long ago as the Pliocene period, which dates from between 5.2 and 1.64 million years ago. Even so, other flightless birds of the Alcidae such as *Mancalla*, *Praemancalla* and *Alcodes* were not closely related to *Pinguinus*. There were indeed smaller flightless alcids (the family to which the great auk belonged) during the Pliocene, which had wings even more penguin-like than those of the great auks. The still-living heirs are the murres and the razorbills.

In the nineteenth century, with the great auk population endangered, a natural disaster brought nemesis closer.

Iceland, geologically speaking, is the youngest part of the Earth, and volcanoes and volcanic eruptions are still extremely active there – the new island of Surtsey, for example, appeared only as the result of the eruption of a submarine volcano in the mid-1960s. Hot springs abound on the mainland, and to walk across a still-warm lava bed is not an unusual experience there. In March 1830, a catastrophe befell the Geirfuglasker: a volcano erupted under the sea, causing a seaquake so great that the contours of the coast were changed, and some skerries and islets disappeared under the waves. The auks would not have been breeding there at that time of year, but they had lost another safe haven, and contemporary records indicate that by 1840 the birds were becoming noticeably rarer. It seems apparent that when they returned to breed, they resettled on Eldey, still hard for men to reach, but much closer to the mainland, easier to land on, and therefore more vulnerable.

Precise numbers are unknown, but as there were fewer birds, their value to collectors increased. Hunting remained intense too, though pickings seem to have been slim: only about fifty were killed

in the years following the disappearance of the Geirfuglasker. It seemed unlikely that the great auk would survive the catastrophes that befell it on Funk Island and on its colony off Iceland. Another problem that it had was the small number of breeding grounds – only eight are known, again based on the evidence of contemporary accounts: Funk Island was by far the largest, but great auk colonies were also identified on the Magdalen Islands, or Bird Rocks, in the Gulf of the St Lawrence River, and possibly off Cape Breton as well; and apart from the Icelandic colonies there were others at St Kilda, on the Faeroes, at Papa Westray in Orkney, and perhaps the Calf of Man. There was also a colony on Penguin Island, off Cape La Hune on the south-west coast of Newfoundland. Captain Taverner, an English sailor, wrote in the early eighteenth century that 'the Penguin Islands [sic] are in the summer time covered with fowle of that name. They are as large as any Tame Goose; their wings are soe small that they can never fly, they get their food by Diving in the Sea.'

The main food was largish fish and crustaceans. Most great auks probably fed near their colonies. However, since the bird was a strong swimmer, it may have hunted as far as 10–30 kilometres away from its base. Apart from the breeding season, the bird was sighted as far as 550 kilometres from land. Its heavy body mass allowed it to forage in a great range of depths, between 75 and a maximum of 150 metres. Most dives would have been shallow, however, since it cost more effort to dive to great depths, and would not have been worth it if food had not been more readily available. It is interesting that gentoo and Adélie penguins, birds of a more or less similar size, dive to similar depths today, and may well reflect the diving patterns – as it were, effort related to food – as the great auks. King penguins (*Apenodytes patagonica*), which weigh around 12 kilograms, can dive to 300 metres.

The auks' colonies disappeared during the eighteenth and early nineteenth centuries. Bird Rocks seems to have had no birds left after 1700, though the French navigator Jacques Cartier mentions capturing several there (he called them *apponatz*) in 1534. They had disappeared from St Kilda by 1760; a last bird was reported killed in the Faeroes in 1808 (though other sources state that they hung on until the 1840s), where there is evidence to suggest that the birds used to be rounded up at sea and driven ashore; and the zoologist Alfred Newton reported in his study of the great auk published in 1861 that the last pair on Papa Westray were destroyed in 1812. Oddly enough, one of the major influences on the great auk's final decline was the war with Napoleon. In 1813, Britain and Denmark were at war, and the British blockade of its ports cut off supplies to the Danes. On the orders of the governor of the Faeroe Islands, a gunboat was sent from there to Iceland with a commission to obtain food by all means possible, and it landed a party on the Geirfuglasker, with a very damaging effect on the great auk population.

Newton believed that the bird had become definitively extinct by 1860. Given that the largest colonies seem to have been those of Funk Island and the Icelandic skerries, themselves representing only modest areas of land, it seems likely that the world populations of great auks were never huge, despite the impression that the numbers breeding in high concentrations locally must have given to sailors. In Tudor times, to give one more example, Richard Hakluyt, the collector and editor of voyagers' tales, reported that Funk Island was 'so exceedingly full of birds that one would think they had been stowed there… In less than halfe an houre we filled two boats full of them, as if they had been stones, so that besides them that we did eat fresh, every ship did powder and salt five or six barrels full of them.'

Migration across the sea and breeding may well have been closely related. Little is known about it, but there is some evidence that gives an informed guess that after breeding there were both southward and northward migrations of adults with their young, at least from colonies in Atlantic Canada. Migrating great auks heading north, for example, during the late summer moved in and out of the Gulf of St Lawrence, where they were observed by George Cartwright in August 1771 near Gray Island. There is no evidence to suggest that great auks were not superlative swimmers: if they had left colonies, for example, off southwest Iceland in early July (after the breeding season) they would have had to swim more than 20 kilometres a day to reach Greenland by September; a prodigious feat, even if aided by the East Greenland Current. However, by comparison, Adélie penguins (*Pygocelis adeliae*), about the same size as far as scientists know as the great auk, can easily make 3 kilometres per hour over long distances.

It could be argued that, despite the depredations, the great auk might still have survived if it had been given the breathing-space necessary to rebuild its population, but the science of extinction was still imperfectly understood 150 years ago because nature's bounty still seemed limitless, and man believed that he could go on abusing it with impunity. Indeed, the concept of imminent extinction was scarcely in even an embryonic stage then. The scientists and natural historians of the time who had become curious about the rare and fascinating great auk may, through their own interest in it, have become the authors of its ultimate doom.

This was the age of the great amateur naturalist, and a rush to collect and classify animal and plant species began. Men such as John James Audubon in America spent a lifetime observing and classifying species (he had a specimen of a great auk in his collection, and reported that they were regularly sighted off the coast of Massachusetts in the winter months during the eighteenth century), and at the same time museums and zoos paid high prices for rare and novel creatures. A study of nineteenth-century auction catalogues presents an astonishing picture of the kind of prices paid.

To give an idea of the relative values, it is worth comparing them with the kind of income enjoyed by a contemporary working man. Trade in the eggs especially began seriously in the 1830s, but became huge towards the end of the century. In 1832, an egg changed hands for £15 15s 6d (about £15.80), at a time when the average annual income for a skilled worker was about £9 10s (£9.50). In 1894, an egg sold for £315, against £83 annual income; and in 1898 the skin of a great auk together with an egg went for £630 – the equivalent of about £50,000 at today's prices. In the early 1970s the remains of a great auk went to a collector in the USA for $30,000. Writing in the nineteenth century, Symington Grieve commented:

> To most people interested in natural history, it would seem that much more interest attached to skins, skeletons, or individual bones of the bird rather than its eggs. The last teach little regarding the habits and structure of the bird compared with others. Yet the prices obtained for eggs are about as high as those obtained for skins, and quite out of all proportion higher than any price obtained for skeletons or bones.

An article in *Punch* magazine on 24 March 1888 describes the sale of an egg: 'A golden egg again – another great auk's egg has just turned up, been put up for aukshun, and knocked down again, without being smashed, fortunately, frail a curiosity as it was to come under the hammer. Mr Stevens of King Street, Covent Garden, sold a very fine egg of the great auk for £225…' Ownership of a great auk's egg conferred status. One major collector had nine of them, and on his death the rumour that the family was about to sell them off caused a short but panicky fluctuation in the highly specialized market. The family was obliged to place an open letter in the press stating that they had no such intention.

For the museums, a dead, stuffed and mounted animal was a prerequisite; in zoos, knowledge of how to care for beasts in captivity scarcely existed, and life behind bars was miserable and short for many animals.

Although it was understood that the great auk had become quite dramatically rare, and the lesson of Funk Island had been noted, there was a refusal or an inability to believe that the great auk as a species was threatened. As late as 1891, a Scottish newspaper reported an alleged sighting of great auks on St Kilda. Perhaps it was wishful thinking. While rarity value made the auk an expensive and attractive prize for collectors and naturalists alike, the general view was that continuing to hunt it down could do no real harm, and nature was always able to replenish itself and there were bound to be hitherto undiscovered auk colonies further north – perhaps in the Arctic. Extraordinarily, the great auk was still listed as alive in some bird guides as late as 1945.

Meanwhile, the Icelanders knew good business when they saw it, but it did not occur to them that very soon they would have finished off the source of their golden egg. They went on hunting the bird as a food source, not just to obtain specimens for scientists. They were later vilified for their actions, but the financial incentives for them to behave as they did were too great for them to ignore, and they were in a sense the simple tools of those truly responsible – collectors and naturalists. And the great auks were not the only seabird whose population declined drastically in the nineteenth century – other species of auk and gannet also reached dangerously low levels of numbers. Murres are still overhunted in Greenland.

The fishermen who set off for Eldey on 3 June 1844 had been hired by a dealer and egg collector from Reykjavik called Carl Siemsen. Siemsen knew the kind of prices he could get for any great auk specimens he could produce, but he did not know the fateful nature of the expedition he had commissioned.

Thirteen years later, in 1857, the great naturalist Johannes Iapetus Steenstrup, Professor of Zoology at Copenhagen University, published his definitive account of the great auk's range. After exhaustive research, he could find no trace of any surviving colonies of the bird in the Arctic regions, or anywhere else. Although ornithologists continued to seek the great auk for several years after that in the hope that a colony might reveal itself, or at least in the hope of a sighting, there never was one, and it had to be acknowledged that the great auk, which had lived alongside us for millennia, was now extinct.

All that remains of the great auk now are seventy-eight mounted skins (mainly from Eldey), twenty-four complete skeletons, two collections of preserved viscera, and around seventy-five eggs.

COMPUTER MODELLING

With the delivery of the live-action backplates, the first production milestone was reached. It meant we'd created the stages upon which each animal's story would unfold. Now it was time to actually build the characters

Constructing six photo-realistic animals was a complex process requiring hundreds of hours' work. Like the building of the animatronics, the work began with detailed research. Once more, zoological researcher Kate Dart gathered together as much visual reference material as possible. In order to model in 3D, visual reference was needed from as many angles as possible, not merely side views, but front, back, top, bottom – precisely those angles that already published pictures or paintings rarely cover. So apart from gathering

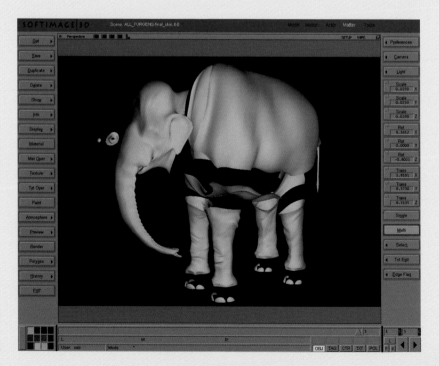

The Columbian mammoth was modelled as separate body parts, including the eye.

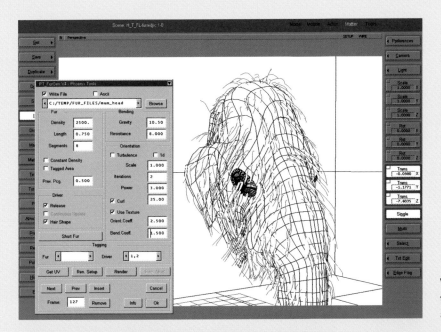

Virtual hairdressing:
the mammoth's hair is
applied one strand at a time.

pictures from a host of books, the research also required visits to zoological collections and museums. We sought out stuffed specimens and skeletons, which were photographed from every angle and in close-up to reference a host of details such as skin, eyes and hair.

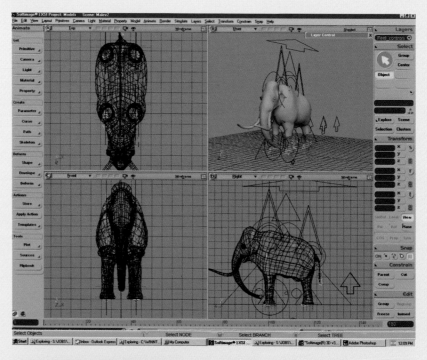

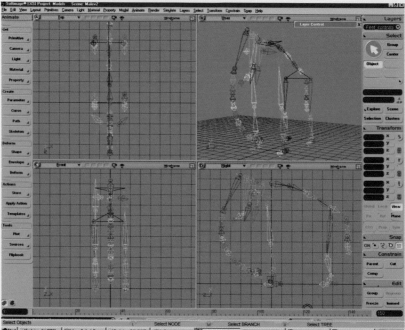

Wireframe model of the mammoth body (top) matched to its interior moving skeleton (bottom).

To start building, the photos were scanned into the computer. For the dodo, modeller Chris Petts used them to build up an accurate composite image of the real bird on top of which he would overlay a computer-generated surface. Each character began life as a series of parts, each of which was modelled separately. The animal was broken down into its constituent parts: head, upper and lower body, thighs, arms (or wings) and so on. Even quite small parts, such as ears and tongues, have to be made separately. Then very simple shapes, called primitives, were created to represent each part – just simple spheres, tubes or cylinders.

Each primitive is a wire-framed 3D surface. The lines that make up the shape can be shaped and reshaped to make each section take on the appearance of the real thing. Slowly, a rough cylinder is sculpted so that it becomes a realistic looking leg. An oval is adapted to become the head, and so on. All through the process the modeller checks what he builds against the reference material.

But getting the shape right involved more than just copying photographs of stuffed specimens and skeletons. We had to be sure that the reference material itself was accurate. This could not always be guaranteed. Stuffed models may look like they are very faithful to

The Tasmanian tiger head.

the real creature, but closer inspection may reveal that they have been wrongly stuffed, creating an inaccurate shape or posture. For example, it became apparent that many models and skeletons of the dodo had posed the bird in a very squat position.

The doubts about their accuracy came from dodo expert Julian Hume. He felt the dodo may have been much more upright, and in the computer we were able to do a small experiment to test this idea. Chris Petts measured available dodo leg bones and used the data to create a movable model of the legs. This enabled us to look at how they articulated. From an anatomical perspective, the upright model moved much more naturally, and showed the leg must have supported the dodo in a much more upright stance than most reconstructions have allowed.

Sometimes, however, the quest for accuracy brought unexpected problems. Our reference source for building the Irish elk's giant antlers was a fossilized set that came from the Dublin Natural History Museum, and we had made a cast of them. The modellers duly went ahead and built an exact 3D copy of the antlers. But when elk expert and programme consultant Dr Adrian Lister saw them, he noticed a bizarre feature he'd never seen before: an additional antler tine.

It turned out to be a pathological deformity unique to the fossil animal we'd taken the cast from. So we had a decision to make. Should we use the deformed antler, because it was exactly the same as a known specimen of the elk? Or should we remove it to make our elk look more normal? In this case our desire to be representative won out over our desire to be absolutely faithful to the reference material. We removed the deformed tine, giving our elk a 'perfect' set of antlers.

Meanwhile, modelling continued building an ever more accurate wire frame model for each part of the creature. Eventually these were put together and for the first time a model of the whole of each animal was created. This was a smooth, grey model – just the basic shape of the creature. Colour, skin texture, feathers and fur were still required.

The first stage in the process is to colour the model as accurately as possible. This means more work on reference material and discussions with experts before final colour approval is given. In addition the colour had to match those already chosen for the animatronic

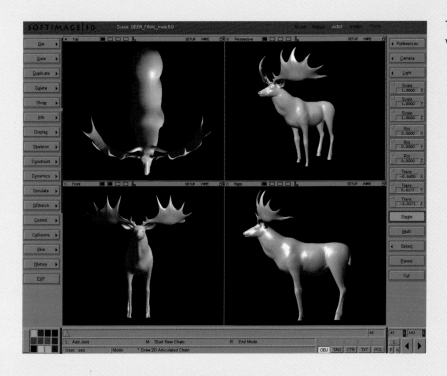

Three-dimensional model
views of the Irish elk.

models. After that, the coloured surface is loaded into a software package that creates fur.

The process of 'virtual hairstyling' now began. Initially, the fur is of uniform length and simply sticks out of the model at right angles. So each hair must then be told in which direction to fall. This is another time-consuming process. Despite being done on a computer, in effect the hair is hand crafted – no small task when you consider a model such as the mammoth may have up to 300,000 hairs. Even then the model still looks a long way from reality.

On real animals, hair is very complex. On the face the hair is short, but gradually lengthens on to the shoulders and neck. The hair flows in different directions and in reality is never entirely uniform. So in the computer, the length and the density of hair across the body has to be adjusted to make the appearance as close to reality as possible.

Having started from a set of photographs on to which 'primitive' computer-generated tubes and cylinders are placed, it took months and hundreds of man hours to create accurate three-dimensional models of our lost animals. Yet they still remained rooted to the spot, unable to move or behave until the final part of the process is completed: computer animation.

TASMANIAN TIGER

Thylacinus cynocephalus

On 13 May 1930, in the Mawbanna district of north-eastern Tasmania, farmer Wilf Batty was eating his lunch when he heard a loud squawking coming from the direction of his chicken coop. Without hesitation, he put down his soup spoon, seized his gun from the wall, pushed a couple of cartridges into the breech, and went out. He imagined the disturbance was being caused by an unwelcome visitor his farmworkers had seen in the area over the past few months.

Once outside, as he had feared, he saw an animal near the coop. As he recalled later, it had had its head under the wire mesh that surrounded the chickens. It looked at him for a moment, and then started to lope away, not particularly quickly, around the back of a shed. Batty followed, though both his kelpies hung back, whimpering.

It was a big, dog-like creature, low-slung, with a large head and a long, stiff tail. Its short brownish fur was banded all across the back with much darker, broad transverse stripes, which gave it its common name. Batty took careful aim as it headed for the perimeter fence, and fired. The animal dropped. Batty walked over and checked that it was dead, and then crossed his yard to see what damage had been done

Tasmanian farmer Wilf Batty with the thylacine he shot in 1930.

to his poultry. The beast he had killed belonged to a type that had a long-established reputation among farmers as a pest, though there had been fewer and fewer of them around over the last twenty years, and this was the first time in Batty's experience that he had heard of one raiding a henhouse. What Batty did not know was that it would also be the last time, for his action had made him the last man to shoot a Tasmanian tiger in the wild. Just over six years later the entire species would be extinct, when the last known survivor died in Hobart's Beaumaris Zoo, of neglect, on 7 September 1936.

Like the dodo, the Tasmanian tiger – or thylacine – has since its demise become a local symbol. (To avoid confusion with the feline tiger, the term 'thylacine' will be used here, following the lead of biologist and veteran 'tiger' enthusiast Eric Guiler – except when quoting contemporary accounts.) 'Thylacine' was the name bestowed on the animal by the urban and scientific communities; in the bush, it had a variety of names, of which 'tiger' was the most common. Among others tried out on it were 'wolf', 'hyena', 'dog-faced dasyurus', 'dog-headed opossum' and even 'zebra opossum'. Today, in a gesture that is sentimental and valedictory, it appears as a supporter of the Tasmanian coat-of-arms and on beer bottle labels.

Action to protect it was not taken until just before the last one died. Scientists today regret that more attention was not paid to the animal by the people who had the privilege of knowing it when it was alive – the best information comes from nineteenth- and early twentieth-century bushmen and trappers. Of the last men who lived at the time when it was still possible to catch the animal, most had died by the mid-1980s, taking their knowledge of it with them. Nevertheless, they left their memories behind. The testimony of such men as George Wainwright, the last 'tiger-man' at the Woolnorth Estate in north-eastern Tasmania, and the Pearce brothers of Derwent Bridge, who between them caught dozens of 'them useless things', survives to both help and tantalize modern researchers. Now, when it is too late, professional and amateur zoologists speculate on the behaviour of an animal lost to us, and argue with each other about who is right and who is wrong, for they have so little to go on. Some – though by now very few – still entertain the hope that somewhere the thylacine may still be alive – but that appears to be wishful thinking.

The thylacine is in a family of its own, the Thylacinadae, of the order Dasyuroidea. Found solely in Australasia, its fossil history goes back as far as the mid-Oligocene – around 25 million years ago. Only in the last 3,000 years of its existence was the thylacine confined to Tasmania: in prehistoric times, this wolf-like marsupial ranged across the Australian continent – the chief predator for two million years. Its wide gape – the jaws opening to almost 100 degrees – made it easily capable of crushing the throat or even the ribcage of the wallabies which, with other smaller mammals such as possums and small kangaroos, were its general prey. In common with other ancient marsupial predators such as the marsupial lion, the thylacine is an example of parallel evolution – the marsupials of Australia (and South America) being long cut off from the far more widespread placental mammals of the rest of the world.

Both marsupial and placental mammals evolved from an animal similar to the Australasian duck-billed platypus, one of the world's oldest and most primitive mammals. About 100 million years ago the two discrete types of mammal went their separate ways, the placentals eventually having a much broader area and variety of environments in which to interbreed and evolve. They give birth to live young that are developed to a stage at which they suckle but are completely separated from the mother. Marsupials have a pouch, a kind of halfway house between the womb and independent existence. The young are born at an embryonic stage and suckle several different mixtures of milk while they are developing, physically separate from, but still totally dependent on, the mother.

In recent human history, marsupials were long considered an inferior, antique mammal. Such an attitude may well have coloured the way in which European settlers in Tasmania regarded the thylacine, utterly unable to understand its true nature. Before those settlers arrived, the thylacine had no natural enemies in Tasmania, surviving in relative harmony with the modest local human population – a few aboriginal tribes had established themselves in Tasmania from about 50,000 to 20,000 years ago.

The first people had arrived on mainland Australia possibly as long as 125,000 years ago, but their real impact on the local animal species did not begin until between 60,000 and – crucially – 40,000

years ago, when the megafauna of the continent began to disappear. By then, it is argued, humans, being intelligent and omnivorous, had developed in numbers and range to a sufficient degree to have invaded all areas occupied by other animals, there to compete with them for the food supply. The large predators, both mammal and reptile, existed in smaller numbers than their prey; in direct competition for it with man-the-hunter, they declined in numbers relatively quickly, to the extent that their populations could no longer reproduce. It is probable that thylacines were not only in competition with the aboriginal tribes of the country for food, but that they were themselves preyed on by man.

Matters were made worse when, about 4,000 years ago, new human arrivals from Asia brought the dog with them – specifically the dingo. Aided by this domesticated hunting collaborator, man's capability became even greater – although in general dogs feared thylacines, and it is likely that their arrival did no more than hasten the marsupial's already inevitable end on the mainland.

Incidentally, the incidence of feral dogs in Australia increased dramatically with the arrival of Europeans, and their destruction of the aboriginal Coorie peoples. Masterless, the dogs ran wild, though aborigines today still keep dingoes as companions and hunting dogs. The dog did not reach Tasmania until 1798 with the arrival of the explorer George Bass. They bred in the company of the sealers who established bases on the island soon afterwards; some ran wild, and some were appropriated by the local aboriginal tribes, who quickly learned how useful they were as the people themselves began an unhappy form of integration with the Europeans. Soon Tasmania would have its own population of feral dogs – whose depredations, as we shall see, were often blamed on the thylacine.

The thylacine was known to the Adnyamathanha people as the *marrukurli*. Oral tradition relegates it to the Dreaming, though with the suggestion that the animal lived on into the not-too-distant past. The last Adnyamathanhan claiming to have seen a thylacine in the bush as a child was a man called Mount Serle Bob, who died in 1919, aged 100. Interestingly, bearing in mind what its reputation would be among European settlers, some aboriginal legends told of thylacines carrying off children to eat them, and Mount Serle Bob warned people to keep their children safely indoors at night, for fear of such attacks.

To this day, reported sightings of the thylacine have continued unabated not only in Tasmania, but on the Australian mainland as well. Nevertheless, no conclusive evidence has been supplied: since the last thylacine died in 1936, there has been no photograph, film or video of one – and no body. Even Eric Guiler, who spent most of the past fifty years on the trail of the vanished animal, nurturing the hope that it might still have survived in the dense interior bush of Tasmania, has now given up. In his study *Thylacine*, published in 1985, he had hoped for traces of a recovered population by the end of the twentieth century. Today he thinks that if they were going to make a recovery, they would have done so by now.

Robert Paddle, a lecturer in psychology at the Australian Catholic University, St Patrick's Campus, has also made a study of historical perceptions of the 'tiger', and published a book about it. Many professional scientists and naturalists disagree with his viewpoints, but his book is challenging, arguing that the Tasmanian tiger is very important to the Australian psyche, which has never been able to reconcile the myth and reality of this creature leading to its being very misunderstood. And misunderstood despite going extinct so recently that there is even film of it – this must be one of the first animals driven to extinction by man to have been recorded on film.

The last thylacine was captured as a cub and taken to Hobart's zoo, where she lived for twelve years, and where the filmed records were made. Nick Mooney, Tasmania's Nature Conservation Branch Wildlife Manager, has studied these films carefully. He notes the breadth of the gape, but also points out the animal's bizarre gait, much more awkward and stiff than a wolf's walk.

Acknowledging that it is hard to draw many (if any) conclusions about the animal's behaviour in the wild from a few reels of film of a highly stressed isolated creature in a zoo, Mooney is interested in comparing the thylacine's skeleton with that of a wolf. The wolf has legs that are not only long, but of equal length. They give the animal a smooth, powerful stride, which makes it a fast runner. The thylacine, on the other hand, does not have the body of a fast runner. Like other marsupials it had proportionately big feet, long hind legs and short front legs, which tended to slow it down.

The wolf had long since become a malign creature in European

myth, and the thylacine, having so many attributes in common (and, as we shall shortly see, also being associated with the 'ferocious' tiger of India), was now to be tarred with the same brush, and for similar reasons: sheep-killer, throat-biter, blood-sucker – all the appurtenances and excuses man invents to cover his own sometimes destructive instincts.

Europeans had reached Tasmania in 1803 as permanent colonizers. The island had probably been discovered by the Dutch explorer Abel Tasman late in 1642, though Portuguese explorers touched there earlier. He named it Van Diemen's Land after his sponsor, Anton van Diemen, Governor of Batavia. Francoys Jacobz, Tasman's pilot-major, led an expedition onshore on 2 December, and reported 'the footing of wild beasts having claws very like a tiger'. This first identification, though it is not of course verifiable, was nevertheless borne out by several subsequent reports by explorers and Dutch East India Company officers during the seventeenth and eighteenth centuries, which mention 'tiger' footprints and sightings. It is possible, however, that up until the end of the eighteenth century the animal encountered or whose tracks were seen was the Tasmanian devil, a scavenging carnivore relative of the thylacine. It was not until the early nineteenth century that an animal which was indisputably a thylacine was recorded.

By this time, one of the cruellest penal colonies then known had been established on the island. Convicts occasionally escaped, but most of them discovered only a tough landscape (like an untamed England in many ways), and no way to get further: they would have had to reach the mainland, and that alone was a difficult task, after which they would have to survive in the vast continent of Australia. That first recorded account of a thylacine comes from a group of escapees, who saw one in the bush. When they returned to prison to face their inevitable flogging and bread-and-water, they described their experience, which was duly noted in the diary of the colony's pastor, Robert Knopwood, on 18 June 1805: 'Am engaged all the morn. upon business examining the 5 prisoners that went into the bush. They informed me that on 2 May when they were in the wood they see a large tyger that the dog they had with them went nearly up to it and when the tyger see the men which were about 100 yards away from it, it went away. I make no doubt but here are many wild animals which we have not yet seen.'

This mild and rational account – which incidentally describes what is generally first-hand accepted thylacine behaviour (given the choice between fleeing and fighting) when confronted with men and at least one dog – did not, alas, stick in people's minds. The image of a 'tyger' did. It is hard to imagine today that even as late as the mid-twentieth century predators were regarded as vicious and aggressive *per se*. Even vegetarian gorillas were regarded as 'savage'.

A couple of months before Knopwood recorded the convicts' report, another possible sighting had been noted by Tasmania's Lieutenant-Governor Paterson, who says that his dogs killed 'an animal of carnivorous and voracious tribe' near Yorktown on the River Tamar. Paterson's description is too vague for us to be sure if the creature the dogs attacked was a thylacine or a Tasmanian devil, but the emotive language in which the animal is described is an early cue for the role in which the thylacine was to be cast. It is interesting too that whereas early illustrations of the thylacine were quite accurate, as time wore on it was presented in an increasingly monster-like light. The mentality of the time dictated that any animal which killed other animals was 'bad'.

Reports of the thylacine such as Paterson's, and most of those that followed, meant that the poor animal was launched on a collision course with man. As new waves of settlers arrived in the first half of the nineteenth century, a kind of hysteria developed. This was an alien place (though Tasmania has often been described as 'little England'), and the fear of dangerous beasts in the hinterland of the undiscovered country was great: out there was an implacable enemy to be exterminated. Just 131 years after the would-be escapees from the penal colony became the first recorded Europeans to see a thylacine, the creature was gone. But what was the connection between the hysteria it generated and its disappearance?

The first question to be answered is, how much of its killer reputation – swiftly and superstitiously ascribed – did it deserve? Zoologist Menna Jones is able to ascertain what neither the old film footage of the last zoo-living thylacine, nor skeletons, can: how it killed its prey.

But one of the problems she faces is that although the animal died out just within living memory, there were no studies conducted of it in the wild. However, its relative, the Tasmanian devil, still lives. The 'devil' (*Sarcophilus ursinus*) is a tough animal, about two-thirds the size of a badger, but capable when the opportunity presents itself of taking new-born lambs – it has a very powerful bite. It is interesting that it did not attract the same hatred that the thylacine aroused, since it looks far more unfamiliar; and the dog-like thylacine was tameable and even occasionally kept as a pet – something that the devil never submitted to.

Young thylacine pups in captivity would play with such objects as a dangled string, just like a kitten or a puppy-dog. Unlike dogs, they did not fight over food. They were generally unaggressive even when grown, and, like many dogs, tolerant of children. Irene Semmens recalled to Robert Paddle that as a child in the 1920s she often played with the children of a family who kept a thylacine as a watchdog. 'If, during the course of a game, a ball landed on or near the thylacine, Irene would just walk up to the animal, pick up the ball and continue playing. The thylacine made no aggressive response… at all. Irene treated it as if it were a well-trained domestic dog kept about the house – and indeed, that was how the animal behaved.'

If threatened themselves, thylacines would elevate their tail and give a warning hiss, which could escalate to a warning growl prior to attack; their usual vocalization was a 'coughing bark'. They could defend themselves well against dogs, if men hunted with them; but of course they were no match for a bullet, and were helpless once caught in a snare. One or two touching stories arise out of man's association with the thylacine in this connection, however. During the Depression of the late 1920s a man called Reg Trigg built himself a bark hut in the Great Western Tiers near the Walls of Jerusalem, where he set about snaring for a living. Finding a young female thylacine in one of his traps, he took her home in a sack and tended her wound. She allowed him to feed her by hand, and at last allowed him to stroke her head, which she seemed to enjoy. He called her Lucy. However, as winter approached, she became restless, and the trapper suspected that a wild male was signalling to her, so he released her. Two years later, he encountered her, waiting for him by one of the tracks he used, with two cubs. Man and beast gazed at each other calmly and affectionately for a few minutes, before she turned unhurriedly and disappeared into the bush with her young. He never saw her again, but by that time it was 1932, and the thylacine race was very close to extinction.

Menna Jones is in the process of making a long-term study of Tasmanian devils, and has used her knowledge to work out how the thylacine may have killed. It lacked the speed of a wolf, so that in order to track down its prey, she argues, it made up for its lack of speed in sheer stamina. It would probably, like devils, have had a relatively low metabolic rate, which would mean that it could maintain a long chase while burning up very little energy.

The thylacine developed a stiff walk, akin to the gait of a dog but less fluid. Tasmanian devils, with far larger hindlimbs than forelimbs, waddle like possums. Thylacines had a superb hunting technique – a mixture of ambush and incredibly persistent pursuit: with a strong sense of smell, they could follow prey for many hours, all day and all night if need be, until the prey gave up and collapsed from sheer exhaustion. The thylacine could not wag its tail and when it turned it did so 'like a ship', but it was agile enough in the bush. Its habit of persistent following (or 'dogging') prey struck fear into men, whom it also latterly sometimes followed, though probably out of curiosity, or possibly in the hope of pickings. Unless severely pressed, a thylacine would never have considered attacking so large, aggressive and unfamiliar an animal as man.

Menna Jones's comparative studies of the jaws of thylacines and wolves enabled her to analyse how the two different animals would have killed. The shape of the canine teeth is different. A wolf's teeth are narrow, and designed to slash. Like its relative the quolls, a thylacine's are oval, and designed to crush, so the Tasmanian tiger would have used its enormous gape to kill with a crushing bite rather than a slashing one. Eric Guiler writes:

I do not know of any written descriptions of the method used by thylacines to kill their prey, but I have talked about it with former hunters who presented me with some degree of agreement on the topic. H. Pearce told me that 'they hunt by lying in wait for their prey and then jump out on it. Or are killed by standing on them and biting through the short rib into the body cavity and ripping the rib cage open.'

Guiler concludes that Pearce's account, corroborated by other similar interviews, 'must carry some weight and although his description of hunting and killing is somewhat different from the general views, the "lying in wait" would fit in with the final phase of the chase.' However, there can be no consensus about the thylacine's hunting methods. For an animal so recently extinguished, little is indisputably known about it, largely because widespread interest in it only began in the last and probably skewed days of its existence: stressed animals whose numbers have been significantly reduced do not behave as they would in normal, healthy circumstances. Indeed, serious study only really began long after the last thylacine had died.

It is known that the thylacine had a delicate appetite, and it was widely believed that it elected to eat only the liver, kidneys, heart and lungs of its victim, along with parts of the soft inner thigh when really hungry, leaving the rest to be scavenged by Tasmanian devils. The fact that it often went for the throat of its victim quickly gave it an association – shared with the wolf – with the vampires and lycanthropes of fairy-tale. It was even credited with living on blood – a concept that was taken seriously for a time. While, as we have seen, some people kept the animals as pets, on the whole the Europeans' fear of the thylacine verged on the irrational. Skulls and pelts would be nailed up on huts as trophies. Even domesticated animals were not immune. One old settler interviewed by Robert Paddle told him that 'fifty years ago Mrs Harrison's brother had a tiger in a cage, at Forrest. It… was quite healthy, but the neighbours were scared of it, and poisoned it after several weeks.' Perhaps they thought it might escape and go after their flocks.

Most Europeans took it for granted that the thylacine was a natural-born killer of sheep, though trappers often said in later interviews that their observations of the animal (and they were the ones who probably came closest to it and were best able to note its habits) ran counter to this. Many gung-ho pioneering stories however deliberately played up the dangers of life in the bush, and the thylacine became a convenient adversary for the intrepid bushman of popular fiction and bar-room boasting. Uninformed 'scientific' disquisitions did not help either.

Soon after Knopwood's and Paterson's reports, Tasmania's Deputy Surveyor-General, George Harris, officially described the new species, and gave it its first scientific name: *Didelphis cynocephala*. This name – 'dog-headed opossum' – was later (in 1834) refined to *Thylacinus cynocephalus*: literally, 'dog-headed pouched dog', though it is normally taken to express the sense of a 'pouched dog with a wolf-like head'.

In 1806 Harris sent his report, with illustrations and 'descriptions from the life' of both the thylacine and the Tasmanian devil, to Sir Joseph Banks, who had accompanied Captain James Cook on his round-the-world trip of 1768–71, and was by now President of the Royal Society. Banks in turn read Harris's account to the Linnaean Society in London, prior to its publication. The description of the

Thylacines in Beaumaris Zoo: (below) a mother and cubs; (opposite) an adult male and female.

thylacine, given here in part, is strikingly accurate, though it adds to the reputation for 'savagery' that man was keen to attribute to the thylacine from the word go:

The length of the animal from the tip of the nose to the end of the tail is 5 feet 10 inches, of which the tail is about 2 feet... Head very large, bearing a near resemblance to the wolf or hyena. Eyes large and full, black, with a nictant membrane, which gives the animal a savage and malicious appearance... Tail much compressed, and tapering to a point... Scrotum pendulous, but partly concealed in a small cavity or pouch in the abdomen.

The whole animal is covered with short smooth hair of a dusky yellowish brown... On the hind part of the back and rump are 16 jet-black transverse stripes, broadest on the back, and gradually tapering downwards...

Only two specimens (both males) have yet been taken. It inhabits amongst caverns and rocks in the deep and almost impenetrable glens in the neighbourhood of the highest mountainous parts of Van Diemen's Land, where it probably preys on the brush Kangaroo, and various small animals that abound in those places.

Europeans first began to settle in Tasmania in the districts around Hobart and the Tamar Valley. Expansion was fast, and by 1820 Hobart was the second-largest town in Australia. In the early days, exploitation of local resources was directed towards sealing and whaling, but the island is temperate, fertile and relatively small, about 67,800 square kilometres in size – slightly smaller than Ireland – and inland surveys quickly opened the way to farming. Agricultural development naturally brought the new settlers into contact with the animals and people already there, and often the new arrivals did not find the original denizens, particularly the predators, convenient.

The land the new arrivals found was a hilly country with open grasslands in its midst, along some parts of the coast, and in the river valleys. Central Tasmania is a low plateau of sedgeland, about 700 metres above sea level. This area was suitable for both herbivorous and carnivorous marsupials. To the west and south are low mountain ranges rising to little more than 1500 metres. These are heavy-duty mountains, however, with deep gorges and thick forest, originally impassable to man. The western ranges and the central plateau have many thousands of lakes that owe their origins to glaciers. The narrow coastal plains, few wider than a couple of hundred metres, also offered great scope to wild animals. Much of the country was covered by woodland, which today has been eroded by logging.

The habitat of the thylacine was altered radically by the Europeans. As early as 1844 large areas of the upper Derwent Valley, the midlands and the east coast had been settled. By 1885 vast swathes of bush had been transformed into farmland, and most of the thriving settler community were farmers of one kind and another – though sheep farming was the main occupation. The sheep, its management, and politics were all to play a key role in the downfall of the thylacine. By 1887 the total sheep population of Tasmania was in the region of 1.5 million.

As Nick Mooney has pointed out, if one studies certain contemporary documents, one gets the impression that the thylacine was a deadly, aggressive and implacable killer. In fact, it provided a perfect scapegoat to cover up bad stock management, over-optimism about the quality of certain terrains to maintain British-bred sheep, marauding of flocks by feral dogs, and rustling. Naturally, presented with a new and easy prey species, some thylacines, especially when their population was in decline, did kill some sheep, but to nothing like the extent their trumped-up reputation suggested, as old trappers and long-time settlers who remembered the animal attested again and again in interviews.

However, sheep farmers needed to cover themselves and used the thylacine as a means of putting pressure on the Hobart government to compensate them for losses – as so little was known about the 'tiger', it was easy to whip up a frenzy of fear and hatred of this secretive animal. Even its name – 'tiger' – suggested a far more awesome animal than the kind of marsupial German shepherd that it actually was. In 1888, a bill was passed offering a £1-per-head bounty on thylacines, an enormous amount in those days, and one that encouraged even more trappers to hunt the animal in its own habitat, far from farms, just to get the money. The impact of this bill, which was not rescinded until 1909, was immediate and devastating. During the period of its imposition over 2,000 animals were killed and, at the peak of the hunting, the government paid a bounty on a 'tiger' every two days. But in the last days, one bounty every year was nearer the mark, so rare had the animal become. It is unlikely that it ever existed in vast numbers, and certainly never to such an extent as to pose an actual threat to sheep farmers' livelihoods. That thylacines were accused of hunting in packs, killing sheep 'for sport', sometimes a hundred in a night, is symptomatic of the hysteria; but far more likely culprits for such depredations when they did occur were packs of feral dogs.

Not all of the early literature on thylacines mentions that it was considered a sheep-killer. Indeed, one report, from 1810, states that settlements were 'free from that destructive animal to Sheep, the Native Dog, the dread of the Stock Holders in New South Wales. The only Animal unknown on the Continent is the hyena opossum [thylacine], but even here they are rarely seen… it flies at the approach of Man, and has not been known to do any Mischief.'

However, between 1803 and 1819 the sheep population of Tasmania grew from thirty to 172,000. In that time two accounts of reports of attacks on sheep by thylacines appeared, both dubious. It is almost certain that the first, telling of an animal that 'was long a

TASMANIAN TIGER: VITAL STATISTICS

Thylacinus cynocephalus

Appearance: large dog- or wolf-like head with wide (100 degree) gape; evenly balanced backbone; forelegs shorter than rear as in all marsupials. Legs shorter than trunk. Deep chest; non-retractable claws. 13–19 dark brown transverse stripes on back and butt of stiff tail. Otherwise brown to light brown dense short fur. Small, erect ears; long muzzle. Oval teeth, designed to crush. Immensely strong jaw.

Size and weight: in wild, may have grown larger than any recorded specimens, but average nose-to-tail-tip length of body about 1.6 metres, of which tail about 54 cm.

Longevity: an animal introduced to London Zoo in 1884 lived for 8 years 5 months. Last thylacine in Beaumaris Zoo, Hobart, died in 1936 aged 12 years – the oldest recorded.

Distribution: only in Tasmania in recent times; widespread especially along the north and east coasts and inland from them. Highest density believed to have been in geographical centre of country. On mainland until about 3,000 years ago, though unsubstantiated Australian sightings persist, as they do in Tasmania.

Reproduction: mating never observed and breeding season uncertain, though traditionally believed to have been around December. Female would produce between one and four young.

terror to the numerous flocks', was no more than an excuse for mismanagement by a notoriously inefficient stockholder, one Edward Lord. However, the myth of the thylacine's voracious ferocity was given even greater currency in a book by W. C. Wentworth published in 1819. It speaks of 'an animal of the panther tribe… which… commits dreadful havoc among the flocks'. Those with an interest in increased investment in the colony of Tasmania were quick to counter this allegation, pointing out how seldom the thylacine was encountered, how rarely it attacked sheep, and how ideal the country was for sheep-breeding and rearing.

The newly formed Van Diemen's Land Company was a consortium of English businessmen that owned large holdings in the north-west of the country. In 1825 they dispatched a new company employee, Edward Curr, who already had experience of sheep farming in the territory, to obtain land and start farming for them. Curr knew his business, and also knew what fine sheep country much of Tasmania was, but he was ill-served by his associates, and he was aware of the inexperience of many stockholders, and that there was a dearth of good shepherds. Few reprieved convicts, or those out on 'good behaviour', or 'ticket-of-leave', had the right kind of experience. English shepherds tended not to be drawn to a life in the colonies.

One of the biggest problems Curr and his colleagues faced was sheep rustling, which had been made a hanging offence in 1813, though the death sentence turned out to be a negligible deterrent. There were also disputes in which a farmer would set his dogs on the sheep of another.

At least Curr knew that there were no really appreciable animal predators on sheep, and advised owners of remoter hill-farms, less likely to be attacked by rustlers or packs of feral dogs, simply to see their sheep bedded down at night, when they 'may be generally certain of finding them in the same place in the morning'.

But then, in 1830, a private bounty scheme was introduced by the Van Diemen's Land Company offering 'rewards for the destruction of noxious animals'. Five shillings was offered for every male 'hyaena', and seven for every female, 'with or without young'. Tasmanian devils and dingoes fetched half the price. But why this abrupt initiative? Had the thylacine population suddenly developed a great appetite for sheep? Or was there another reason?

All was not going well on the new company settlements: convict-labourers mutinied, unskilled indentured servants from Ireland were unhappy and hopeless workers, poor supplies left men short of food, and clearing and fencing programmes were not being carried out on schedule. Curr's associates and managers disagreed among themselves and showed little efficiency in carrying out their duties. Some of the land purchased by them on the company's behalf was unsuitable for sheep, as Curr discovered when he visited the holdings. But the company's directors in London had to be appeased. As early as 1828 and 1829 cattle and then sheep were released into land not ready for them, with no shelters built, and in consequence of mismanagement many hundreds of animals died. The fate of the venture was not helped when the normally benign weather turned icy, and the worst winter for many years dragged on through 1829. Nor did it abate, but continued through the latter months of the year – lambing-time in those latitudes – on and on until Christmas.

Curr needed excuses and he needed a scapegoat. Even though Van Diemen's Land Company records show that predation by feral dogs was a greater problem than predation by thylacines ever was, Robert Paddle has argued convincingly that the thylacine suited Curr's purposes perfectly. And Eric Guiler points out that 'It would not have been practical to have introduced a bounty scheme for dogs, as the canine population of the state would have been endangered, including useful working animals such as cattle and sheep dogs as well as household pets.'

Though the local press rumbled about Curr's inefficiency, London would take him at his word – he had proved himself to be an experienced Tasmania hand to them, and if he bought time by blaming exotic predators, he might be able to turn the whole thing round yet. In any case, the directors needed to cover themselves and have a story to fob the shareholders off with. In Tasmania, the water was muddied by locals' inability always to distinguish between thylacines and devils anyway, but there remained few genuine reports of sheep predation (though devils certainly scavenged the carcasses of already-dead animals), and in 1831 the bounty for both sexes of

thylacine was raised to ten shillings, an appreciable amount at that time, and enough of an incentive for men to go off deliberately to hunt down the 'vermin', even deep in the bush, far from any farm. Curr needed a bigger quota of caught animals than he had hitherto obtained to show to his London masters.

Despite the fact that more and more local people were beginning to identify feral dogs as a palpable menace to their sheep, somehow the thylacine was kept centre stage as the leading villain. This was very largely because it continued to be portrayed in print as such. Thylacines were described as voracious predators 'prowl[ing] from the mountains' and descending in 'considerable numbers' on sheep folds. The emotional and lurid language employed in such descriptions fuelled people's imaginations, and was useful to Curr, who stuck to his original scapegoat.

By the early 1840s the Tasmanian economy was in a mess. Cheap convict labour was no longer available to private companies, there had been three years of failed wheat-harvests, and there was a heavy drought in north-west Tasmania, where the best Van Diemen's Land Company holdings were. It seemed an odd time to introduce a renewed thylacine bounty scheme, though this time at only six shillings per head, but that is what was done, and not only did the company run one, but other private enterprises too, for greater or shorter periods. In the meantime, so few thylacines were being encountered that the so-called 'tiger-men' on large holdings – trappers employed specifically to snare the animal, and earn their bread by that activity – could not find enough to support themselves. As early as 1835 one such tiger-man employed at the company's large,

150,000-acre Woolnorth holding, was contracted to work for a regular ration of flour and half-a-guinea (about 53 pence) for every thylacine he killed. The money was supposed to cover all his other expenses, but he could not begin to earn his keep and had to be employed on a shepherd's wage in order to maintain him. No thylacine at all was seen at Woolnorth between 1836 and 1839, though there was a resurgence later.

As the century wore on, the war against the thylacine did not relent. 1884 saw the setting up of local residents' action groups, such as the Buckland and Spring Bay Tiger and Eagle Extermination Society. Some newspapers supported the campaigns, though others printed letters and articles from farmers pointing out that sheep had far more to fear from devils and wild dogs than eagles and thylacines. The final straw came in 1888 when, after two years of lobbying by a local politician called John Lyne, the Hobart government introduced a state bounty by a majority of one vote. The Chief Inspector of Sheep was appalled at such an intemperate move. By that time too there was a considerable counter-movement that supported the nascent idea of conservation, and it had substantial support. Lyne, however, was an ambitious politician who had been member of parliament for the district of Glamorgan since 1879. Given his association with the Glamorgan Stock Protection Association, he saw strong political advantage in making himself the hero of the major sheep-owners' faction. Juggling figures, flying in the face of counter-arguments that pointed to many other reasons for the decline in sheep-farming's fortunes, such as a glut of wool on the international market, drought, and the fact that rabbits had run wild, bred alarmingly, and were competing with sheep for grazing, he doggedly pressed his case until he won his point. He had no interest in the thylacine *per se* at all, and simply used it to further his career. Ironically, he retired from politics, a victim of deafness, in 1893, only five years after his coup.

The government passed the bill, but it soon found that it had miscalculated badly. Few people wanted to see rich farmers subsidized to the extent of the £500 allocated for the destruction of thylacines, at £1 per animal in the first year. In the event, not much of the money was spent – trappers were already being offered five or six times that amount by zoos, museums and private collectors. Eric

Guiler mentions that in 1888, seventy-two adult thylacines and nine juveniles were presented for government bounty. As we have seen, the figure then hovered around the one hundred mark until 1905, when it first halved, and then fell off to fifteen and two in 1908 and 1909 – the last year of the bounty. At its height, between 1899 and 1901, it rose to between 132 and 140 adult animals a year.

The thylacine was long held to be a solitary hunter, but recent evidence has suggested that although it was never a pack animal like the dog, it was known to hunt in small groups. Certain trappers have recalled occasions when one thylacine would drive prey towards an ambush laid by another. It was an intelligent, relatively sophisticated, probably social animal, and the erosion of its numbers quickly led to stress, affecting normal behaviour, and giving rise to breeding difficulties.

After mating, and a gestation period of a month, up to four pups would be born. They would then crawl to the nipples located within the pouch, which was backward facing to protect the pups inside from branches and twigs in the undergrowth that would have brushed against the mother as she moved around, but also so that she could eject the pups if highly stressed and needing to flee. It is thought that thylacines could mate again in the same year if they lost a litter.

The young would leave the pouch three or four months later, staying with the mother until she next came into season. It is thought that a female might have taken up to two years to produce a maximum of four young. Eric Guiler writes: 'Several authors have claimed that thylacines have a lair or other retreat and that the young are brought up there, the lair being variously described as a hollow log, hollow tree, a cave or rock cavity… There is no doubt that a thylacine will have its favourite sleeping places but this is very different from establishing a den in which the young are brought up.'

Solitary thylacines – unmated individuals – tended to have no fixed abode, using their great stamina to range far and wide. While the precise nature of the thylacine's social grouping is unknown, Robert Paddle presents the following argument:

Parallel evolution, for example, of marsupial moles and placental moles, clearly illustrates that the evolutionary construction of an organism and its behaviour is an interactive product of information in the environment together with information in the genes... In recognition of this, the parallel evolution of the marsupial wolf [thylacine] and placental wolf should come as no particular surprise, nor, in parallel with physical similarity, should surprise be expressed at social similarity. In placental wolves, while multiple monogamous pairings are known within the one extended family pack, the most usual expression of reproductive behaviour – both in the wild and in captivity – is of a single breeding pair within a nuclear family group or extended family pack... The incidence of monogamy within the Mammalia reaches its peak within the Canidae (the dog family)... It should come as no surprise that the commonest group structure of the marsupial wolf reflects a similar family orientation. It was, indeed, this very social group structure that made thylacines such effective, responsive and rewarding pets.

It has to be added, however, that this view of great sociability in thylacines is not shared by some professional scientists, who point out that there is no group living among related dasyures such as the Tasmanian devil. An animal needs good practical reasons to be social and the thylacine, some argue, would not have reaped any great benefits from group living.

As hunters, thylacines, equipped with elliptical pupils, hunted at night as individuals, to increase the element of surprise; but when hunting in groups, they needed visual contact with one another, and so used daylight. Most of the animals on which they preyed were at their most active at dusk, but would doze well after dawn.

As mentioned earlier, the thylacine's principal prey species were (small) kangaroos, wallabies and possums. Its diet also included wombats and smaller mammals including, latterly, the imported rabbit, as well as birds such as ducks and teal. Thylacines, unlike Tasmanian devils, were never natural scavengers, and although some took scraps from campsites, outside captivity they generally were not known to eat anything they had not themselves killed. Predation on farmed imported species, such as cattle, goats and pigs, seldom occurred (only one instance of an attack on a goat and one on a pig

'Benjamin', the last known living thylacine in Beaumaris Zoo.

have been specifically recorded), and predation on sheep was uncommon. In captivity, a thylacine would take and eat a chicken. In 1921, a man called Harry Burrell took a series of photographs of a thylacine in a private zoo in Tasmania doing precisely that. As the evidence of the chicken-wire enclosure is indistinct in the photographs, some took them to be of an animal in the wild, and that led to tales of thylacines raiding hen-houses. It is possible that there were a few such raids, but there cannot have been many, for by then the thylacine itself was on the brink of extinction.

As the thylacine became rarer, some naturalists and even some farmers became more interested in it. Hitherto, as we have seen, it had endured a very bad press indeed – described as 'stupid' and 'primitive', with the implication that it was all in all a dull creature hardly worth bothering about – and this on top of its vilification and 'monsterization' as a pest and relentless sheep-killer. In addition to any scientific or even affectionate interest (from those who made pets of it), collectors wanted trophy pelts of the world's top marsupial carnivore. Prices for thylacine skins rose as the animal became scarce, and museums and zoos worldwide waved their chequebooks in search of specimens. Prices for live animals rose to £68 paid in 1911 by London Zoo, for example.

The first live specimens went on display at Regent's Park in 1850, and one was also on display in a menagerie in Hobart from 1854. By the 1860s they were represented in zoos internationally. They were scarcely seen in the wild at all, but they were still regarded by many people as useless, retarded and uninteresting creatures.

Robert Paddle is angered by the fact that scientists, 'who should have known better, failed this animal so badly'. He adds: 'There was an Australian school of thought at the time that argued that marsupials were simply a doomed species in every way inferior to placental mammals. Evolutionists belonging to this school expounded the idea that the Tasmanian tiger was an evolutionary dead-end, a joke creature not worth preserving; and as the tiger's numbers dwindled, they put almost no pressure on the then government to stop the killings.'

In any case, in 1906, eighteen years into the open season on thylacines, a strong warning of its decline manifested itself: numbers of pelts brought in by trappers halved in comparison with the previous year. Eric Guiler argues that a critical point had been reached:

> It's certainly doubtful that trappers alone could have wiped out such a huge proportion of tigers in such a short time. If an animal is hunted to extinction, its numbers drop gradually, and then in the places where it was hunted most extensively first. What we had with the Tasmanian tiger was a dramatic drop all over the island. I do think that the impact of the bounty left the tiger very vulnerable: with a decreasing population the animals would have found it increasingly difficult to mate. Dwindling numbers could have led to inbreeding, which in turn would have left the thylacine genetically depressed and more exposed to disease.

After 2,207 kills claimed, the Hobart government had to pay out only once in the scheme's final year, 1909, the year in which, ironically the Chair of Zoology was established at the University of Tasmania. Guiler believes that one possible reason for such a dramatic decline in the early years of the last century may indeed have been disease: 'By 1900 there were intriguing reports from the trappers saying that they'd found listless thylacines in their snares which made little attempt to free themselves. Before this point it was highly unusual to find a passive tiger in a snare – so much so that trappers were officially warned to be cautious of the ensnared animal.'

The only reasonable explanation for this shift in behaviour is illness. At the time, other marsupials, Tasmanian devils for example, but also possums, succumbed to a dire strain of either canine distemper or pleural pneumonia. Affected animals exhibited hair loss, diarrhoea, and scabbing all over their bodies. It is possible, though there

are no records to prove it, that the thylacine was similarly affected. In 1908 the species had suddenly become rare, and if they had simply been hunted to extinction then, as Guiler has stated, logically it would follow that they would disappear first from the places where hunting them was most vigorous; but this was not the case. The French family of Tin Pot Marsh went on happily trapping and shooting thylacines from the mid-1880s through to about 1922.

It is possible that the female thylacine shot by Wilf Batty in 1930 was an animal in season but unable to find a mate. To men like Batty the animal was still a pest, worthy of extermination, though cynicism also played a part in its demise. The female may have been weakened by disease, and desperate for food, which induced her to attempt her fateful attack on Batty's poultry.

It was not until August 1929 that the Animals and Birds Protection Board of Tasmania passed a motion affording partial protection to the thylacine by pronouncing a closed season on them for December, which was repeated the following year – December was believed to be the breeding month for the animal. In September 1933 a sub-committee was formed to consider further methods of protection, but hunting permits continued to be issued until 1936.

The Tasmanian government gave the thylacine legal protection on 14 July 1936. Less than two months later, the last one known died in Beaumaris Zoo. The zoo itself, under the care of an inexperienced and uninterested director, was run-down and neglected, and the animal had been badly looked after. The door connecting its cage to its sheltered night-quarters was frequently left locked, leaving it at the mercy of the elements. It was not fed regularly. Early September brought below-freezing temperatures at night alternating with blistering highs of 38 degrees Celsius during the day. The tree in its compound which was the thylacine's only hope of shade had already shed its leaves for the winter.

It was enough to ensure her demise. Ironically, it was only after her death that Tasmania, believing the animal still existed, began to think about setting up sanctuaries for it. Now there is talk of reintroducing the thylacine by cloning.

COMPUTER ANIMATION

With the models complete, the final and most important phase of the production process began. Many hours' work had gone into making the animals appear accurate. But they were still just statues. If they couldn't be made to walk and move convincingly, we'd never be able to get close to the illusion of reality.

More than that, they also had to be placed against the backplates we shot on location. Only then would they appear to move around and behave in the real world. Every nuance of lighting, shadow and angle would have to be adjusted to achieve this realistically. Also, in many scenes they'd have to interact with actors, be reflected in water, brush through vegetation or swim under water, processes aided by the use of blue-screen shooting on location.

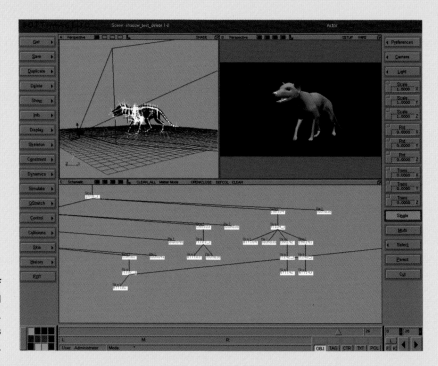

Wireframe model of Tasmanian tiger (left) and greyscale version (right). The chart (below) shows layers of animation.

	CG SEQUENCE 5 *TIGER KILLS WALLABY*	
5.1 CG 8"	Top wide shot valley tiny CG tiger chasing prey in far distance, disappears over horizon	
5.2 Rushes & CG 3"	WS wallaby crossing frame, followed by CG tiger	
5.3 CG 1"	MS CG tiger running past camera	
5.4 Rushes 2"	CU wallaby turning direction	
5.5 Rushes 4"	WS wallaby bounding in forest	

First stage storyboard of the Tasmanian tiger feeding scene.

Second stage storyboard.
A simple line drawing is
added to the backplates
shot in Oregon.

Knowing how the animals would move was a primary concern from the very start of production. Before computer modelling, animatronic construction or backplate shooting, we drew up storyboards to help visualize the way in which the creature would behave in each scene. Our plan was to replicate all the normal ingredients of a wildlife documentary. To do this, we'd need to create a lot of animal behaviour – basic actions like running, walking and swimming, but also hunting scenes, mating rituals, feeding, nest building, egg laying, herding, and looking after young. After that they'd have to be woven together to tell the story of how scientists believed each species had actually become extinct. This meant translating events that overcame an entire species into a plausible scenario that would see the death of a single individual. Here accuracy was vital, and so was powerful storytelling.

In some cases we had particular historical events to re-create – the deaths of the very last great auks and Tasmanian tigers. In other cases we had to rely more on imagination. For example, we knew there was ample evidence that sabre-toothed tigers had become trapped in the tar pits of La Brea, a fact that gave us an actual death scenario to re-create. But, of course, the La Brea tar pits didn't lead to the extinction of the whole sabre-toothed species. Climatic change over a relatively long period, and over a wide geographic range, had a hand in that. So our problem was how to bring these two elements together.

In the end, director Jenny Ash came up with a storyboard for a dramatic final scene in which the female tiger, weakened and starving as a result of environmental changes, would perish in the tar pits, attracted by the prospect of food. We felt a scene like this neatly brought the two threads together while remaining scientifically plausible.

With the first rough storyboards then drawn up, we were able to see how much variety of action would be required for each animal. We knew the kind of environments in which it would take place and we knew what kind of camera angles would be needed for the backplates. It also allowed us to see which shots and movements would be simple for computer animators to achieve and which more complex. This enabled us to allocate resources as efficiently as possible.

On location, the backplates were shot according to the action drawn up in the first-stage storyboards. The animatronics were shot at the same time to match lighting and background. Next came the process of putting the two together. More storyboards were now needed, this time incorporating the backplate imagery. Now we could work out the precise choreography of the animation in the shot. At first the position of the animals was simply drawn in roughly. Later on, the finished CG models were added. Now the animators knew what they had to achieve and animation began.

As in hand-drawn animation, the principle is simple: move the image tiny amounts, frame by frame, to give the illusion of movement. However, in reality, making each animal move was a complex task. To begin with, the models themselves have nothing in them that's movable. They are no more than hollow computer-generated shapes. In order to move them they need to be joined to some kind of internal framework. This required the building of what's known as a rig or skeleton.

The rig is a simple 'stick figure' model of the animal. Every limb and joint is represented. For example, legs must have a thigh and a calf, joined by a knee joint, and be attached to the body by a hip joint and to the foot by an ankle. Some parts are more complex. A long neck or a trunk need multiple joints to give them the necessary flexibility and control.

The rig is the basic tool with which to animate the computer model. The way in which the real creature would move, what scientists call the gait, then has to be programmed in. To do this, each joint must be assigned movement parameters. These tell the computer in which direction, and how far, each joint can move in relation to another. This controls factors such as how far the head can move from side to side, the length of each step, and the side-to-side rock of the pelvis. Getting the movement parameters right for each joint affects the reality of the overall movement. It's a complex business. On a simple walk, all the joints must be moving in relation to each other – if any one doesn't do so correctly, the accuracy of the creature's walk will be affected. This means many hours' work go into making sure the rig is set up correctly, and our academic consultants were brought in to verify that the computer-generated locomotion

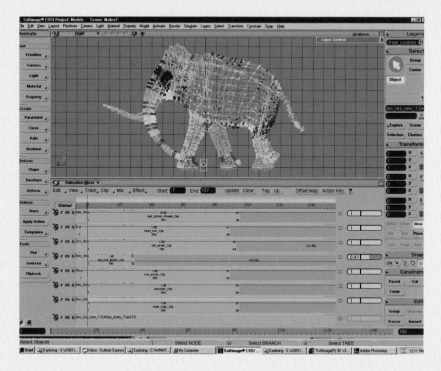

Wireframe model of the mammoth, showing stages in a walking sequence.

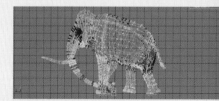

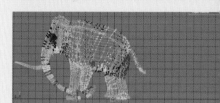

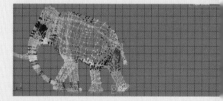

matched their view of the real animals' movement, derived from studying their skeletons.

Next, each rig was programmed to replicate a wide range of movements. We needed a whole repertoire such as walks, runs, head turns, tail swishing, ear flicks and eye blinks. These basic moves can then be combined and recombined, speeded up or slowed down, to give many hundreds of different behaviours.

Finally, the rig then had to be joined to the 3D model – a process known as enveloping. This links the static model to the movable rig; when it is programmed to move, so does the model. Enveloping is another process that sounds simple, but in practice is very painstaking work. Because the model was originally built up out of separate units, when the rig is programmed to move, sometimes tears appear between these different elements. For example, the leg may split from the body when it walks. Often it requires many hours to iron out these bugs so the animal can walk and move exactly as required.

When the process is complete, we finally had fully movable characters. These were composited into the backplates, the necessary lighting and shading added, and each frame rendered to produce the finished TV image. The creatures had been brought back to life.

BIBLIOGRAPHY

INTRODUCTION

Da Vinci, Leonardo (manuscript held in Leicester Codex)

Hooke, Robert, *Discourses on Earthquakes* (London, 1705)

Raup,David, *Extinction: Bad Genes or Bad Luck?* (W.W. Norton & Co, New York, 1992)

COLUMBIAN MAMMOTH

Carrington, Richard, *Mermaids and Mastodons* (Chatto and Windus, London, 1957)

Lister, Adrian and Bahn, Paul, *Mammoths* (Marshall, London, 1994)

Mol, Dick, Agenbroad, Larry D. and Mead, Jim I., *Mammoths* (The Mammoth Site of Hot Springs, South Dakota)

www.amnh.org/science/biodiversity/extinction/Day 1

www.cpluhna.nau.edn/Biota/megafauna_extinctions

www.discovery.com/exp/mammoth/discoveries

SABRE-TOOTHED TIGER

Turner, Alan, *The Big Cat and Their Fossil Relatives* (Columbia University Press, New York, 1997)

IRISH ELK

Barnosky, Anthony D., '"Big Game" Extinction Caused by Late Pleistocene Climate Change: Irish Elk in Ireland' (*Quaternary Research*, 25, 1986)

Gould, Stephen Jay, 'The Origin and Function of "Bizarre" Structures: Antler Size and Skull Size in the "Irish Elk"' (*Evolution*, 28, June 1974)

Hayden, Tom, 'On the Trail of "The Giant Irish Deer"' (*Wild Ireland*, Sept/Oct 2000)

Kitchener, A.C., Bacon, G.E. and J.F.V. Vincent, 'Orientation in Antler Bone and the Expected Stress Distribution, Studied by Neutron Diffraction' (*Biomimetics*, Vol 2 No 4, 1994)

DODO

Kitchener, Andrew C., 'On the external appearance of the dodo' (*Archives of Natural History*, 20, 1993)

Livezey, Bradley C., 'An ecomorphological review of the dodo (Raphus cucullatus) and solitaire (Perzophaps solitaria), flightless Columbiformes of the Mascarene Islands' (*Zoological* Society, London, 230, 1993)

Van Wissen, Ben (ed.), 'Dodo - exhibition catalogue to accompany The Fate of the Dodo' (Zoölogisch Museum, Universiteit van Amsterdam, 1995)

GREAT AUK

Birkhead, T.R., *Great Auk Islands: A Field Biologist in the Arctic* (Poyser, London, 1993)

Bourne, W.R.P., 'The Story of the Great Auk' (Archives of *Natural History Archives*, 20)

Fuller, Erroll, *The Great Auk* (privately printed)

The Audubon Society Book of Water Birds (Harry N. Abrams, NYC, 1987)

www.rom.on.ca/biodiversity/auk/aukhist

TASMANIAN TIGER

Guiler, E.R., *Thylacine* (Oxford University Press, Melbourne, 1985)

Paddle, Robert, *The Last Tasmanian Tiger* (Cambridge University Press, Cambridge, 2000)

INDEX